THE CARTOON GUIDE TO THE

THE CARTOON GUIDE TO THE ENVIRONMENT

Larry Gonick and Alice Outwater

HarperPerennial

A Division of HarperCollins*Publishers*

FIRST EDITION

Library of Congress Cataloging-in-Publication Data

Gonick, Larry.
        The cartoon guide to the environment / Larry Gonick and Alice Outwater. — 1st ed.
             p.   cm.
        Includes bibliographical references and index.
        ISBN 0-06-273274-9
        1. Environmental sciences—Caricatures and cartoons. 2. Environmental education—Caricatures and cartoons. I. Outwater, Alice B. II. Title.
        GE105.G66   1996   95-49891
        363.7—dc20

96 97 98 99 00 RRD 10 9 8 7 6 5 4 3 2

TO BOB, WITHOUT WHOM THIS
BOOK WOULD NEVER HAVE BEEN
WRITTEN

◆CONTENTS◆

Acknowledgments

WE WOULD LIKE TO THANK CAROL COHEN, OUR EDITOR AT HARPERCOLLINS, FOR HER UNFAILING SUPPORT AND GOOD HUMOR, AND OUR AGENT, VICKY BIJUR, NOT ONLY FOR HANDLING THE BUSINESS DETAILS BUT ALSO FOR VOLUNTEERING TO PLAY CERTAIN OTHER ESSENTIAL ROLES, INCLUDING BABY SITTING(!). AND THANKS TO JOSEPH MONTEBELLO, ART DIRECTOR AT HARPER, FOR GRACIOUSLY HANDLING THE UNCONVENTIONAL ASSIGNMENT (FOR THE BOOK PUBLISHING INDUSTRY) OF DEALING WITH AN ILLUSTRATOR/ AUTHOR WHO WAS TOO BUSY WORKING ON THE "GUTS" TO TURN IN HIS COVER ART ON TIME.

PROFESSORS PENNY CHISHOLM AND ADEL SAROFIM REVIEWED THE MANUSCRIPT AND OFFERED MANY USEFUL SUGGESTIONS, BOTH IN WRITING AND CONVERSATION. WE ARE PARTICULARLY GRATEFUL TO CHISHOLM FOR KEEPING MUM HER PLANETARY PESSIMISM UNTIL AFTER THE MANUSCRIPT WAS COMPLETE. CARL WUNSCH OFFERED INSIGHTS INTO OCEAN CIRCULATION AND KATE KIRBY HELPED WITH THE GAS.

JOHN DOWNER TURNED GONICK'S HAND LETTERING INTO THE CUSTOM TYPEFACE AVAR. MICHAEL CASTLEMAN MADE INVALUABLE SUGGESTIONS REGARDING ACKNOWLEDGMENTS.

THANKS TO THE KNIGHT SCIENCE JOURNALISM FELLOWSHIP PROGRAM FOR ENABLING THE CARTOONIST TO SPEND A YEAR AT MIT, WHERE THE INTELLECTUAL JUICES GUSH IN TORRENTS, AND TO VICTOR McELHENY, THE PROGRAM'S DIRECTOR, FOR ENCOURAGING THE FELLOWS TO PURSUE THEIR INDEPENDENT COURSES, WHICH IN THIS CASE MEANT FINISHING *THE CARTOON GUIDE TO THE ENVIRONMENT*.

AND SPECIAL THANKS TO ANNEMARIE RAFTERY FOR TAKING CARE OF SAM WHILE HIS PARENTS WERE BUSY, AND TO BOB AND THERESA TRACY, KIND NEIGHBORS.

·CHAPTER 1·

FORESTS AND WATER

OUR STORY BEGINS IN A PLACE THAT'S BEEN CALLED THE MOST FAR-FLUNG INHABITED ISLAND IN THE WORLD: *EASTER ISLAND,* A 64-SQUARE-MILE SPECK IN THE PACIFIC OCEAN, 2300 MILES FROM ANYWHERE.

"HOW INAPPROPRIATE TO CALL THIS PLANET EARTH, WHEN CLEARLY IT IS OCEAN."

—ARTHUR C. CLARKE

REMOTE, BUT NOT **DESERTED...** FROM TIME TO TIME VISITORS DROPPED BY... LIKE THE DUTCH ADMIRAL **ROGGEVEEN** IN 1722. ARRIVING ON **EASTER SUNDAY,** HE NAMED THE ISLAND AFTER THE DATE OF ARRIVAL, AND LEFT THE FIRST WRITTEN ACCOUNT OF THE PLACE AND THE PEOPLE WHO LIVE THERE.

ACCORDING TO ROGGEVEEN AND OTHER 18TH-CENTURY REPORTS, SOME 3000 ISLANDERS EKED OUT A WRETCHED EXISTENCE BY FARMING BANANAS, SUGAR CANE, AND SWEET POTATOES FROM POOR, ROCKY SOIL. THE ONLY FRESH WATER CAME FROM MURKY LAKES INSIDE VOLCANIC CRATERS. THERE WAS SCARCELY A TREE ON THE ISLAND, AND THE PEOPLE WERE "SMALL, LEAN, TIMID, AND MISERABLE."

BUT AMIDST THE SQUALOR
WERE SOME **SURPRISES...**
ESPECIALLY SOME 800
*MASSIVE STONE
STATUES* SCATTERED
ACROSS THE ISLAND,
SHOULDER TO SHOULDER,
THEIR BACKS TO THE SEA.
HOW WERE THEY CARVED?
HOW QUARRIED? HOW
MOVED? HOW ERECTED?
AND BY WHOM?

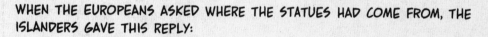

WHEN THE EUROPEANS ASKED WHERE THE STATUES HAD COME FROM, THE
ISLANDERS GAVE THIS REPLY:

FAILING TO RECOGNIZE SARCASM WHEN THEY HEARD IT, THE EUROPEANS EMBARKED ON A SERIES OF WILD SPECULATIONS THAT HAVE CONTINUED TO THE PRESENT DAY: **SPACEMEN** SET UP THE STATUES WITH **ANTI-GRAVITY DEVICES**... A HIGHLY CIVILIZED **LOST CONTINENT** HAD SUNKEN INTO THE SEA, LEAVING ONLY EASTER ISLAND BEHIND... THEY WERE FLUNG INTO PLACE IN ONE PIECE BY **VOLCANIC ERUPTIONS**, ETC. ETC. ETC.

SINCE THE ISLANDERS HAD EITHER FORGOTTEN WHAT HAPPENED OR DIDN'T FEEL LIKE SHARING, IT WAS LEFT TO WESTERN SCIENTISTS AND HISTORIANS TO PIECE TOGETHER THE STORY WITH *CALIPERS, SHOVELS, MICRO-SCOPES,* AND *ETHNO-GRAPHIC SURVEYS.*

AND HERE IS WHAT THEY FOUND OUT...

4

AROUND THE YEAR 400, EASTER ISLAND WAS COLONIZED BY **POLYNESIANS.** VARIOUS FEATURES OF THE ISLANDERS' SKULLS, THEIR BLOOD TYPES, SOCIAL SYSTEM, LANGUAGE, AND CROPS ARE ALL POLYNESIAN.

THE FIRST THING IS — DO SOMETHING ABOUT ALL THESE TREES!

POLLEN SAMPLES TAKEN FROM LAKE BEDS SHOW THAT THE ISLAND WAS THEN THICKLY COVERED WITH VEGETATION. HACKING OUT CLEARINGS FROM THE JUNGLE, THE POLYNESIANS BUILT AND PLANTED, AND SOON THEY ENJOYED A TYPICAL POLYNESIAN DIET OF YAM, TARO, BREADFRUIT, BANANA, SUGAR, COCONUT, CHICKEN, AND POLYNESIAN RAT (SMALL AND TASTY!).

AH... PARADISE... IN HARMONY WITH NATURE... THROW ANOTHER LOG ON THE FIRE... AND PASS THE RAT...

THEIR LIFE WAS RICH... THEIR BABIES THRIVED... THEY POPULATED THE ISLAND WITH LITTLE EFFORT... AND IN THEIR COPIOUS SPARE TIME, THEY CARVED STONE MONUMENTS, ESPECIALLY **STATUES.**

5

THE STATUES WERE MOVED AND SET UP WITH LOGS AND ROPE. RECENT EXPERIMENTS* HAVE PROVED THAT WHEN A STATUE IS STRAPPED UPRIGHT ONTO LOGS CARVED INTO SLED RUNNERS, AND RAISED ONTO A TRACK OF WOODEN ROLLERS, A COUPLE OF DOZEN PEOPLE COULD PULL IT EASILY!

THE PROBLEM IS STOPPING IT !!

*BY AMERICAN GEOLOGIST CHARLES LOVE.

SO THEY CUT DOWN A LOT OF TREES, NOT JUST FOR ROLLING STATUES, BUT ALSO FOR FIREWOOD AND BUILDING MATERIAL...

LOOKIN' GOOD!

AND BY THE YEAR 1400, THERE WAS SCARCELY A TREE LEFT STANDING ON EASTER ISLAND...

OOPS!

AND WHAT GOOD ARE TREES?? YOU MIGHT ASK... READ ON...

THE WATER CYCLE

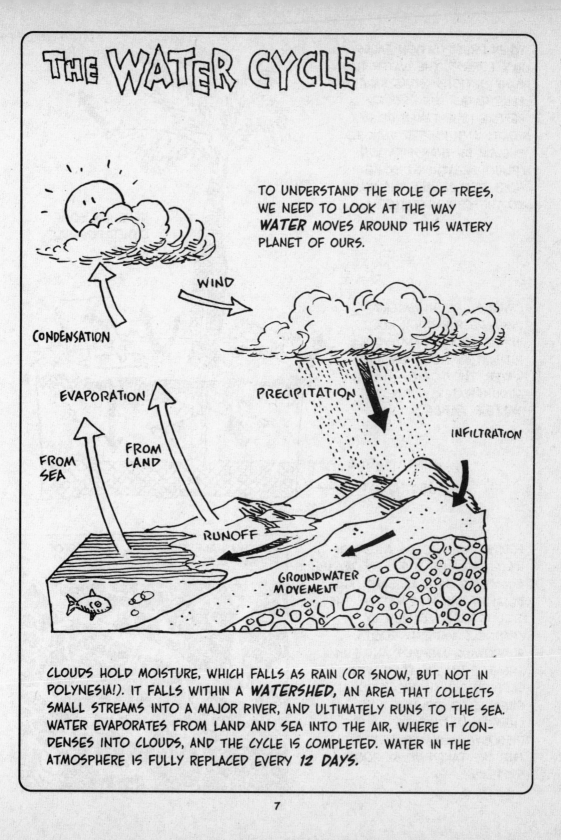

TO UNDERSTAND THE ROLE OF TREES, WE NEED TO LOOK AT THE WAY **WATER** MOVES AROUND THIS WATERY PLANET OF OURS.

CONDENSATION

WIND

EVAPORATION

PRECIPITATION

INFILTRATION

FROM SEA

FROM LAND

RUNOFF

GROUNDWATER MOVEMENT

CLOUDS HOLD MOISTURE, WHICH FALLS AS RAIN (OR SNOW, BUT NOT IN POLYNESIA!). IT FALLS WITHIN A **WATERSHED**, AN AREA THAT COLLECTS SMALL STREAMS INTO A MAJOR RIVER, AND ULTIMATELY RUNS TO THE SEA. WATER EVAPORATES FROM LAND AND SEA INTO THE AIR, WHERE IT CONDENSES INTO CLOUDS, AND THE CYCLE IS COMPLETED. WATER IN THE ATMOSPHERE IS FULLY REPLACED EVERY **12 DAYS**.

WHEN PRECIPITATION FALLS IN A FOREST, THE WATER HAS MANY OPTIONS: SOME BARELY PENETRATES THE GROUND BEFORE IT IS TAKEN UP BY ROOTS AND PASSED BACK TO THE AIR BY TRANSPIRATION (PLANT BREATHING). SOME GOES DEEPER, ALL THE WAY TO THE GROUNDWATER.

HERE ARE SOME OTHER POSSIBLE PATHS:

(WHAT'S GROUNDWATER? DIG A DEEP ENOUGH HOLE IN THE GROUND, AND YOU'LL HIT WATER. THAT'S GROUND-WATER. THE TOP OF THE GROUNDWATER IS THE **WATER TABLE**.)

WATER TABLE ⟶
GROUNDWATER ⟶

FOREST SOIL HOLDS A LOT OF WATER BECAUSE IT'S SO **POROUS**: A MIXTURE OF CLAY, SAND, AND DECAYING ORGANIC MATTER, THE SOIL IS HONEYCOMBED BY CHANNELS MADE BY ROOTS, BURROWING ANIMALS, AND FUNGI. THE TOP LAYERS TEEM WITH BACTERIA, WHICH BREAK DOWN ORGANIC COMPOUNDS INTO CHEMICAL NUTRIENTS THAT DISSOLVE IN WATER, DRIP DOWN, AND ARE TAKEN UP BY ROOT SYSTEMS.

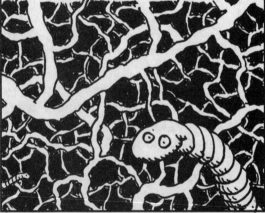

AROUND HALF THE TOTAL VOLUME OF FOREST SOIL IS EMPTY SPACE!

BY CONTRAST, OPEN LAND HAS LESS BIOLOGICAL ACTIVITY AND SO HOLDS LESS WATER.

SO THE PLANTS AND ANIMALS HOLD MORE!

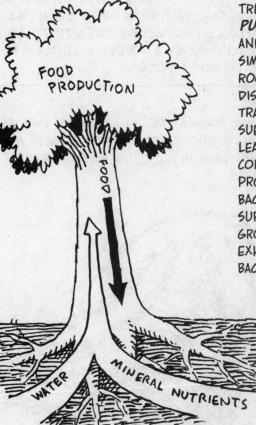

FOOD PRODUCTION

FOOD

WATER

MINERAL NUTRIENTS

TREES ARE GIANT **WATER PUMPS,** MADE UP OF ROOTS AND LEAVES CONNECTED BY A SIMPLE PLUMBING SYSTEM. THE ROOTS SUCK UP WATER AND DISSOLVED MINERALS, WHICH TRAVEL UPWARD THROUGH TISSUE BENEATH THE BARK TO THE LEAVES, WHERE THEY ARE CONVERTED INTO SUGAR AND PROTEINS. THESE FOODS TRAVEL BACK DOWN TO THE ROOTS TO SUPPORT THE ROOTS' FURTHER GROWTH. THE LEAVES ALSO EXHALE A LOT OF WATER VAPOR BACK INTO THE ATMOSPHERE.

UNDERGROUND, HUNDREDS OF MILES OF TREE ROOTS ANCHOR THE TREE TO THE GROUND AND HOLD THE SOIL IN PLACE. MARVELOUS THING, A TREE!

FALLING LEAVES MAKE UP A GOOD PART OF THE DEAD ORGANIC MATTER ON THE GROUND. SHADE REDUCES EVAPORATION FROM THIS MATERIAL, AND SO, IN A SENSE, THE TREE *MANUFACTURES THE SOIL IN WHICH IT GROWS.*

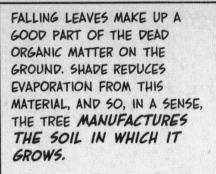

I AM GREEN WITH GRATITUDE!

MADE IN THE SHADE!

THIS TAKES TIME, OF COURSE. FIRST, SMALL PLANTS PRODUCE A LITTLE TOPSOIL, WHERE BIGGER PLANTS CAN GROW. EVENTUALLY, ENOUGH NUTRIENT-RICH EARTH BUILDS UP TO SUPPORT A FOREST, WITH ALL ITS TEEMING SPECIES.

WHAT HAPPENED ON EASTER ISLAND? WHEN A TREE IS CUT DOWN AND ITS ROOTS DIE, THE TOPSOIL LOSES ITS ANCHOR. 4 TO 5 FEET OF EARTH MAY EVENTUALLY BE WASHED AWAY, AND THERE IS NO QUICK WAY TO REPLACE IT.

WITHOUT FORESTS TO ABSORB RAIN AND REPLENISH GROUNDWATER, THE ISLAND'S STREAMS AND SPRINGS DRIED UP... THE AIR BECAME LESS HUMID, AND RAINFALL DIMINISHED. AS FERTILE TOPSOIL ERODED, CROP YIELDS FELL... THERE WAS NO WOOD FOR HOUSES... NO FIBERS FOR FISHING NETS OR SAILCLOTH, NO LOGS FOR CANOES...

AND WORST OF ALL: WITHOUT CANOES, NO *ESCAPE!*

RIVALRY FOR RESOURCES LED TO PERMANENT WARFARE. UNFORTUNATELY, CLAN PRESTIGE WAS DISPLAYED BY *ERECTING STATUES,* SO THE LAST TREES WERE PROBABLY CUT DOWN IN A FRENZIED EFFORT TO SHOW OFF...

SO THERE!!

THE POPULATION PEAKED AT AROUND 7000 PEOPLE IN 1550 AND COLLAPSED SO QUICKLY THAT SOME 400 STATUES WERE LEFT UNFINISHED IN THE QUARRIES.

WHAT ARE THEY DOING NOW?

REDUCING THE POPULATION FURTHER...

BUT THE RIVAL CLANS FOUGHT ON... THEY PULLED EACH OTHERS' STATUES DOWN... AND BY 1860, EVERY STATUE ON THE ISLAND HAD BEEN TOPPLED.

WHEW!

11

THE POINT OF THIS STORY IS NOT THAT THE PEOPLE OF EASTER ISLAND WERE SOMEHOW STRANGE, SILLY, OR DIFFERENT FROM ANYONE ELSE. QUITE THE CONTRARY: LIKE THE REST OF US, THEY WERE **CREATURES OF HABIT**, AND THEIR WAY OF LIFE— FARMING, FORESTRY, BUILDING, AND DISPLAY—WAS HARD TO CHANGE.

EASTER ISLAND IS VERY SMALL. FROM ITS SUMMIT YOU CAN SEE THE WHOLE THING. THE PERSON WHO CUT DOWN THE LAST TREE MUST HAVE KNOWN THERE WERE NO MORE—AND STILL HE CUT IT DOWN.

OUR PLANET, WHILE MUCH LARGER, IS STILL FINITE. LIKE THE ISLANDERS, WE TOO CAN SEE IT ALL, AND LIKE THE ISLANDERS WE HAVE NO MEANS OF ESCAPE. IS IT STILL POSSIBLE THAT WE CAN TAKE STOCK OF OUR RESOURCES AND CHANGE OUR HABITS IN TIME TO AVOID THE FATE OF EASTER ISLAND?

«CHAPTER 2»

MORE CYCLES

THE WATER CYCLE, OR *HYDROLOGIC CYCLE* AS
SCIENTISTS CALL IT, IS ONLY ONE OF THE CHEMICAL
CYCLES ENGINEERED BY LIFE. THE EARTH'S ELEMENTS
CONSTANTLY MOVE AROUND IN ENORMOUS,
INTERLOCKING LOOPS, CREATED AND DIRECTED
BY LIVING ORGANISMS.

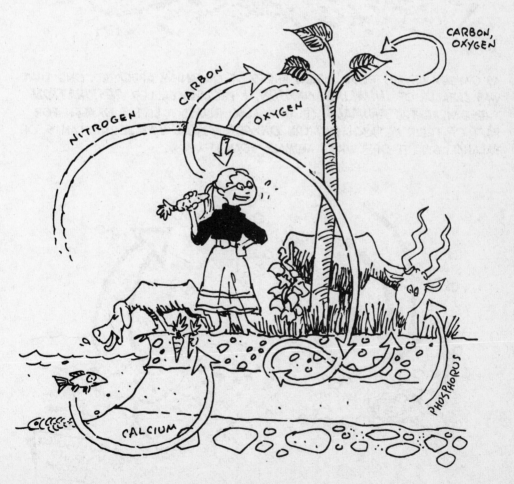

TAKE **OXYGEN**, FOR EXAMPLE... WHEN THE EARTH WAS YOUNG, SOME 4 BILLION YEARS AGO, THERE WAS LITTLE FREE OXYGEN IN THE ATMOSPHERE. THEN LIFE APPEARED, AND SOON THERE WERE MICROBES CALLED **CYANOBACTERIA**, THEN LARGER CREATURES CALLED **PLANTS**, THAT EXHALED PURE MOLECULAR OXYGEN (O_2) AS A WASTE BY-PRODUCT.

AS OXYGEN LEVELS ROSE, A NEW BREED OF ORGANISM APPEARED, ONE THAT WAS CAPABLE OF **INHALING OXYGEN**, A PROCESS CALLED **RESPIRATION**. THESE WERE THE **ANIMALS**. (PLANTS ALSO INHALE A LITTLE OXYGEN FOR PART OF THEIR METABOLISM.) THE **OXYGEN CYCLE** WAS BORN: PLANTS, ON BALANCE, GIVE IT OFF, WHILE ANIMALS SOAK IT UP.

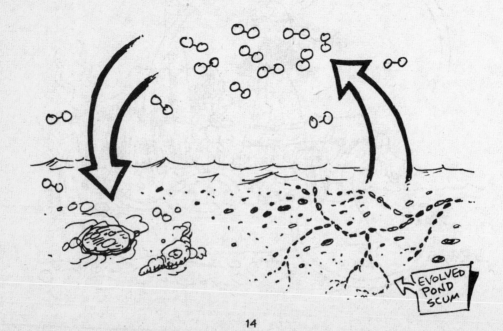

ANIMAL POPULATIONS SWELLED TO CONSUME THE ABUNDANT OXYGEN, UNTIL A LIMIT WAS REACHED, AND THE WHOLE SYSTEM CAME INTO **DYNAMIC BALANCE.** FOR THE PAST 2 BILLION YEARS, PLANTS AND ANIMALS HAVE HELD THE OXYGEN LEVEL AT A VERY STEADY **21%** OF THE TOTAL ATMOSPHERE.

IMPORTANT NOTE: AS IT HAPPENS, ANIMALS EXHALE **CARBON DIOXIDE GAS** (CO_2), WHICH PLANTS ABSORB AS A RAW MATERIAL FOR BUILDING THEIR CARBON-BASED TISSUES. PLANTS ARE CALLED

PRODUCERS

BECAUSE THEY MAKE ("PRODUCE") ORGANIC MATTER DIRECTLY FROM CO_2 AND OTHER CHEMICALS.

WE CREATE ALL LIVING MATTER, AND WHAT DO WE GET FOR IT?

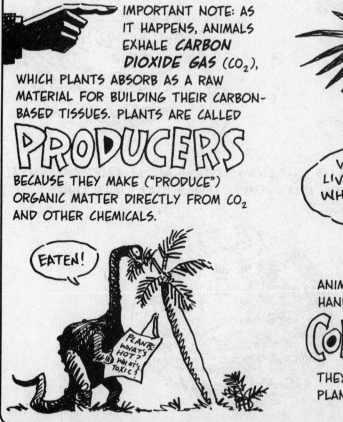

EATEN!

PLANTS: WHAT'S HOT? WHAT'S TOXIC?

ANIMALS, ON THE OTHER HAND, ARE

CONSUMERS:

THEY GET FOOD BY EATING PLANTS OR OTHER ANIMALS!

ONE INTERPRETATION OF THE OXYGEN STORY IS THAT THE **TOTALITY OF LIFE ON EARTH REGULATED THE ATMOSPHERE:** LIFE CREATED AND MAINTAINED CONDITIONS FAVORABLE TO THE GROWTH OF MORE LIFE. COULD THIS BE TRUE OF OTHER CYCLING CHEMICALS AS WELL?

BELIEVE IT!

IN THE 1970S, THE BRITISH CHEMIST JAMES LOVELOCK AND AMERICAN BIOLOGIST LYNN MARGULIS PROPOSED THE **GAIA** (GUY-UH) **HYPOTHESIS:** THEY SUGGEST THAT THE WORLD IS AN INTERCONNECTED, BIOLOGICAL BEING, WHOM THE SCIENTISTS CALLED GAIA AFTER THE GREEK EARTH GODDESS.

LOVE YOUR GREEN OUTFIT, GAIA!

THE GAIA HYPOTHESIS SAYS THAT LIVING SYSTEMS ACT AS **ENVIRONMENTAL REGULATORS,** ENSURING THAT CHEMICAL LEVELS STAY WITHIN LIMITS CONGENIAL TO GAIA'S HEALTH.

I'M BIG! I'M STRONG! I CAN TAKE CARE OF MYSELF!

MESS WITH ME AT YOUR PERIL!

FOR ANOTHER EXAMPLE, **SALT** IS CONTINUOUSLY WASHED FROM LAND TO SEA, YET THE LEVEL OF OCEAN SALINITY HAS BARELY CHANGED SINCE LIFE BEGAN. LOVELOCK SUGGESTS THAT CORAL REEFS EXTRACT SALT FROM THE OCEAN BY WALLING OFF **EVAPORATION POOLS**, WHERE SALT IS ISOLATED FROM THE SEA.

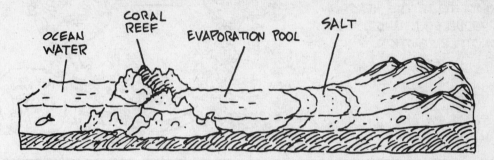

IN LOVELOCK AND MARGULIS' VIEW, THE ABIOTIC (NONLIVING) WORLD—THE **ATMOSPHERE** (GAS), **HYDROSPHERE** (WATER), AND **GEOSPHERE** (THE HARD STUFF)—IS REGULATED BY THE **BIOSPHERE** (THE LIVING WORLD). THE **ECOSPHERE** IS THE INTERACTION BETWEEN THE BIOSPHERE AND ALL THE WORLD'S ELEMENTS.

THE GAIA THEORY SAYS THAT WHENEVER AN ABIOTIC CONDITION BEGINS TO GET OUT OF HAND, LIFE RESPONDS IN SOME WAY THAT PULLS THE CONDITION BACK UNDER CONTROL, DAMPING DOWN THE FLUCTUATION.

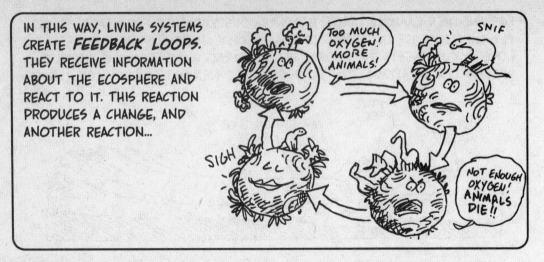

IN THIS WAY, LIVING SYSTEMS CREATE **FEEDBACK LOOPS.** THEY RECEIVE INFORMATION ABOUT THE ECOSPHERE AND REACT TO IT. THIS REACTION PRODUCES A CHANGE, AND ANOTHER REACTION...

LOVELOCK AND MARGULIS BELIEVE (OR HOPE) THAT THESE FEEDBACK LOOPS ARE MAINLY STABILIZING AND REGULATORY. A CONDITION MAINTAINED IN THIS WAY IS CALLED **HOMEOSTASIS,** A **DYNAMIC STEADY STATE,** FULL OF FLUX AND CHANGE, BUT ULTIMATELY TUNED TO OPTIMIZE CONDITIONS FOR LIFE.

THIS VERY ATTRACTIVE IDEA CONTAINS A GOOD DEAL OF TRUTH... BUT AS WE'LL SEE LATER, FEEDBACK LOOPS CAN ALSO BE **DESTABILIZING,** AND THE GAIA HYPOTHESIS REMAINS CONTROVERSIAL.

BUT TRUE OR FALSE, THE GAIA HYPOTHESIS REMINDS US TO THINK OF THE BIOSPHERE AS A SINGLE GIGANTIC SYSTEM.

OF THE 90+ NATURALLY OCCURRING ELEMENTS ON EARTH, ABOUT 40 ARE REQUIRED BY LIVING ORGANISMS. SINCE VIRTUALLY NO NUTRIENTS COME FROM OUTER SPACE, THESE ATOMS ARE CONSTANTLY RECYCLED. A CARBON ATOM ON YOUR NOSE MAY HAVE ONCE BEEN A TRICERATOPS' TOENAIL.

THE **MACRONUTRIENTS**—THOSE NUTRIENTS WE NEED IN LARGE QUANTITIES—ARE CARBON, OXYGEN, HYDROGEN, NITROGEN, PHOSPHORUS, SULFUR, CALCIUM, AND POTASSIUM.. THESE ELEMENTS MAKE UP MORE THAN 95% OF THE MASS OF ALL ORGANISMS, AND THEY MOVE AROUND A LOT. FOR EXAMPLE:

1. CARBON ATOM ENTERS BLOODSTREAM AND BECOMES PART OF TOENAIL.

4. CARBON ATOM GOES THROUGH MORE CHANGES, LANDS ON YOUR NOSE.

2. TOENAIL FALLS OFF AND IS EATEN BY BACTERIA.

3. BACTERIA ARE STIRRED UP BY WIND, RAIN, VOLCANOES.

MICRONUTRIENTS INCLUDE ZINC, IRON, COPPER, CHLORINE, AND IODINE.

WHEN A CREATURE DIES AND IS BURIED IN SEDIMENTS, ITS ELEMENTS SINK BELOW GROUND AND ARE LOST FOR EONS—UNLESS SCAVENGERS EXTRACT KEY ELEMENTS FROM THE CORPSE AND SPREAD THEM AROUND.

JUST DOING GAIA'S DELICIOUS DIRTY WORK!

UGH. AND WHO EATS DEAD VULTURES?

BACTERIA GOT **NO** TASTE...

IF ORGANIC MATTER DOES SINK UNDERGROUND, IT USUALLY RECYCLES VERY SLOWLY. IT MAY NOT COME UP AGAIN FOR MILLIONS OF YEARS, UNTIL EROSION OR GEOLOGICAL UPHEAVAL EXPOSES IT.

OR UNTIL WE EXTRACT THE SUCKER!

GASEOUS ELEMENTS—OXYGEN, NITROGEN, AND CARBON (IN CARBON DIOXIDE)--CYCLE RAPIDLY, BECAUSE THE AIR IS FULL OF THEM AND BLOWS THEM AROUND. THEY MAY RECYCLE INTO LIFE AND OUT AGAIN WITHIN HOURS OR DAYS.

IT'S SO EASY TO MOVE UP HERE!

BETWEEN THESE EXTREMES ARE MEDIUM-TERM CYCLES OF ELEMENTS CIRCULATING THROUGH THE HYDROSPHERE AND OTHER LIVING ORGANISMS.

IN SHORT, THE RATE OF BIOGEO-CHEMICAL RECYCLING DEPENDS LARGELY ON THE ACCESSIBILITY OF THE ELEMENT'S MAJOR RESERVOIR.

FAST

MEDIUM

SLOW

EACH ELEMENT HAS A CYCLE, BUT WE DO NOT FULLY UNDERSTAND HOW ALL OF THEM ARE REGULATED.

THERE'S SO MUCH WE DON'T KNOW ABOUT YOU!

AND YOU CALL YOURSELVES "SAPIENS!"

BEST KNOWN ARE THE THREE BIG ONES:

NITROGEN

(AN ESSENTIAL INGREDIENT OF PROTEINS AND DNA): ALMOST 80% OF THE ATMOSPHERE IS PURE NITROGEN (N_2), A FORM WHICH IS USELESS TO MOST ORGANISMS.

ATMOSPHERIC NITROGEN

NH_3
NO_3

FIXED NITROGEN IS TAKEN UP AND USED BY OTHER PLANTS, WHICH MAY IN TURN BE EATEN BY ANIMALS.

BUT CERTAIN BACTERIA, ESPECIALLY THOSE LIVING ON ROOT NODULES OF PEAS, BEANS, ALFALFA, ALDER TREES, ETC., CAN FIX NITROGEN, I.E., CONVERT IT TO NITRATES (NO_3) AND AMMONIA (NH_3).

EXCRETED, TOO!

CARBON,

THE BASIC BUILDING BLOCK OF ALL LIFE:

CARBON IS FOUND IN THE AIR, AS CARBON DIOXIDE, CO_2, IN WATER, AS DISSOLVED CO_2, AND IN THE GROUND, AS CARBONATE ROCKS (SODA, LIMESTONE, CHALK) AND OTHER COMPOUNDS.

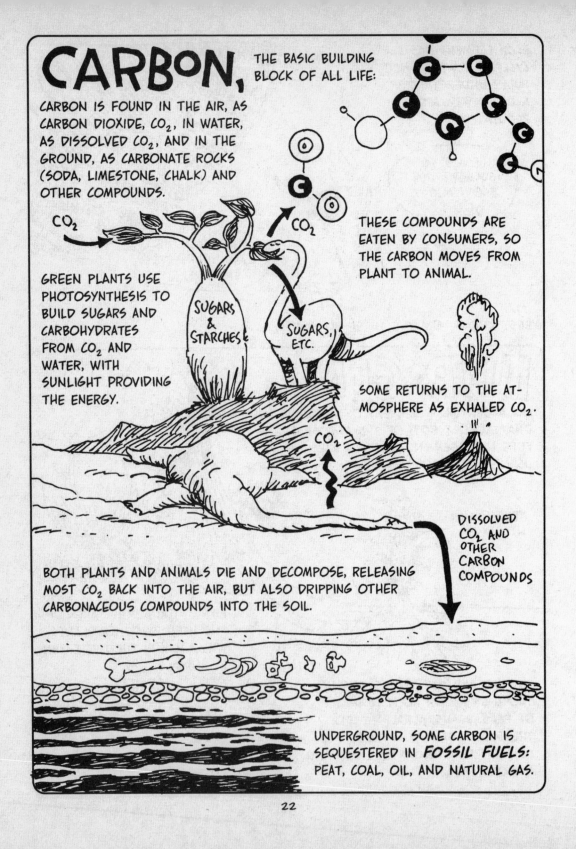

CO_2

CO_2

GREEN PLANTS USE PHOTOSYNTHESIS TO BUILD SUGARS AND CARBOHYDRATES FROM CO_2 AND WATER, WITH SUNLIGHT PROVIDING THE ENERGY.

SUGARS & STARCHES

SUGARS, ETC.

THESE COMPOUNDS ARE EATEN BY CONSUMERS, SO THE CARBON MOVES FROM PLANT TO ANIMAL.

SOME RETURNS TO THE ATMOSPHERE AS EXHALED CO_2.

CO_2

DISSOLVED CO_2 AND OTHER CARBON COMPOUNDS

BOTH PLANTS AND ANIMALS DIE AND DECOMPOSE, RELEASING MOST CO_2 BACK INTO THE AIR, BUT ALSO DRIPPING OTHER CARBONACEOUS COMPOUNDS INTO THE SOIL.

UNDERGROUND, SOME CARBON IS SEQUESTERED IN *FOSSIL FUELS*: PEAT, COAL, OIL, AND NATURAL GAS.

PHOSPHORUS

IS A KEY ELEMENT IN METABOLIC ENERGY TRANSFERS, BONES AND TEETH, AND OTHER TISSUES AND MOLECULES. IT MOVES IN A *SEDIMENTARY* CYCLE.

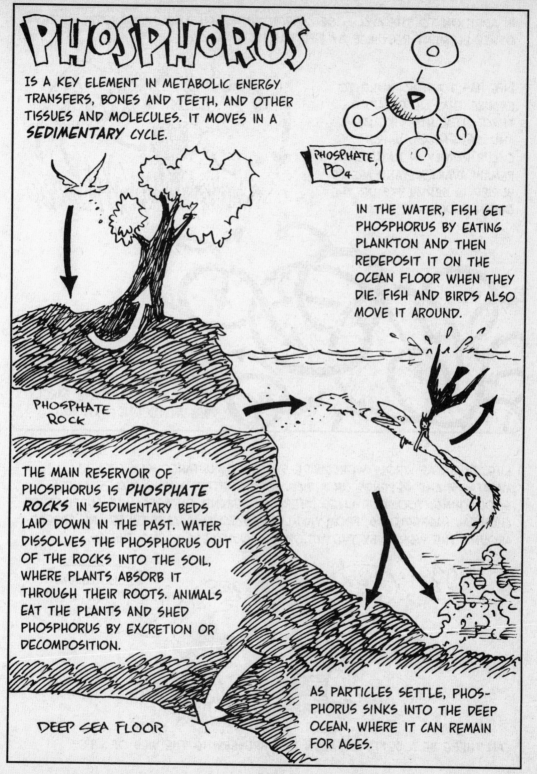

PHOSPHATE, PO_4

IN THE WATER, FISH GET PHOSPHORUS BY EATING PLANKTON AND THEN REDEPOSIT IT ON THE OCEAN FLOOR WHEN THEY DIE. FISH AND BIRDS ALSO MOVE IT AROUND.

PHOSPHATE ROCK

THE MAIN RESERVOIR OF PHOSPHORUS IS *PHOSPHATE ROCKS* IN SEDIMENTARY BEDS LAID DOWN IN THE PAST. WATER DISSOLVES THE PHOSPHORUS OUT OF THE ROCKS INTO THE SOIL, WHERE PLANTS ABSORB IT THROUGH THEIR ROOTS. ANIMALS EAT THE PLANTS AND SHED PHOSPHORUS BY EXCRETION OR DECOMPOSITION.

DEEP SEA FLOOR

AS PARTICLES SETTLE, PHOSPHORUS SINKS INTO THE DEEP OCEAN, WHERE IT CAN REMAIN FOR AGES.

IN ADDITION TO THE CYCLES OF CARBON, NITROGEN, AND PHOSPHORUS, EVERY OTHER ELEMENT REQUIRED BY LIFE HAS ITS OWN CYCLE.

LIFE HAS TO WORK HARD TO ENSURE THAT NECESSARY TRACE ELEMENTS LIKE *ZINC* AND *SELENIUM*, WHICH OCCUR RARELY IN NATURE, REMAIN AVAILABLE AND AREN'T BURIED IN SEDIMENTS OR THE OCEAN FLOOR.

ALL THESE CYCLES INTERACT WITH EACH OTHER IN COMPLICATED WAYS WE HAVE BARELY BEGUN TO UNDERSTAND!!

LIFE, TAKEN AS WHOLE WORLDWIDE SYSTEM, SIMULTANEOUSLY CREATES, MAINTAINS, AND DEPENDS ON A RICH SET OF INTERLOCKING CYCLES. AND A GOOD THING, TOO! OUR LIVES LITERALLY DEPEND ON ADEQUATE SUPPLIES OF CHEMICAL INGREDIENTS, FROM AMINO ACIDS TO ZINC, AND THESE ARE SPREAD AROUND THE WORLD BY THE ACTION OF COUNTLESS LIVING THINGS.

☆@# BIRD!!

NO...YOU SHOULD SAY, "THANKS FOR THE PHOSPHORUS!"

CAN THERE BE A BETTER REASON FOR PRESERVING THE WEB OF LIFE?

· CHAPTER 3 ·

EVOLVING SYSTEMS, STRUGGLING INDIVIDUALS

UNDER THE INFLUENCE OF LIFE, THE EARTH'S CHEMICAL CYCLES WERE CREATED, MODIFIED, AND MAINTAINED OVER THE EONS. COMPLEX WEBS OF ORGANISMS MOVE PRECIOUS TRACE ELEMENTS AROUND. SOME HAVE EVEN ARGUED THAT THE EARTH BEHAVES LIKE A GIANT LIVING THING, REGULATING CHEMICAL CYCLES TO ITS OWN ADVANTAGE.

AT LEAST, THAT'S HOW IT LOOKS FROM A DISTANCE, WHEN BIO-, GEO-, AND HYDROCHEMICAL CYCLES ARE VIEWED AS *SYSTEMS.* FROM THE STANDPOINT OF THE *INDIVIDUAL ORGANISM,* HOWEVER, THINGS LOOK A LITTLE DIFFERENT. BORN INTO THIS VAST FLUX OF CYCLING CHEMICAL RESOURCES, THE INDIVIDUAL HAS A SIMPLER, MORE SELFISH GOAL:

THE INDIVIDUAL ORGANISM STRIVES TO **EAT** AND **AVOID BEING EATEN** LONG ENOUGH TO **REPRODUCE SUCCESSFULLY** (I.E., SO THE OFFSPRING SURVIVE).

WELL, IT'S A START, ISN'T IT?

ALTHOUGH THE INDIVIDUAL IS PART OF THE SYSTEM, THE INDIVIDUAL'S NEEDS ARE NOT NECESSARILY THE SAME AS THE SYSTEM'S NEEDS. FOR EXAMPLE, AN EASTER ISLANDER NEEDED ONLY **ONE TREE AT A TIME** FOR FUEL OR BUILDING MATERIAL, WHILE THE SOCIAL SYSTEM NEEDED A **FOREST** TO RUN SUCCESSFULLY.

AGAIN: WE SET UP STATUES TO INCREASE CLAN PRESTIGE TO IMPROVE ACCESS TO RESOURCES SO YOU CAN REPRODUCE MORE SUCCESS- FULLY.

THANK US.

INDIVIDUALS ARE BORN INTO A WEB OF SYSTEMS— FAMILY, COMMUNITY, SPECIES, BIOSPHERE—THAT CREATE OPPORTUNITIES AND IMPOSE LIMITS. AT THE SAME TIME, THESE SYSTEMS ARE CREATED BY THE ACTION OF ALL THEIR MYRIAD INDIVIDUALS.

A REAL CHICKEN-AND-EGG SITUATION!

IN THE INTERPLAY OF INDIVIDUAL AND SYSTEM,

EVOLUTION TAKES PLACE.

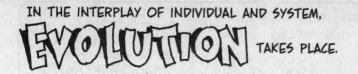

INDIVIDUALS COMPETE WITH EACH OTHER FOR RESOURCES. EVERY CREATURE TRIES TO SUSTAIN ITS OWN LIFE LONG ENOUGH TO REPRODUCE, AND IF THAT MEANS SOMEBODY ELSE GOES HUNGRY, TOO BAD!

THAT IS SO ENVIABLE.

TIGER SALAMANDER LARVA EATING ITS COMPETITOR

THIS DOESN'T MEAN THAT CREATURES DON'T COOPERATE ALSO! SOMETIMES COOPER-ATING WITH OTHERS IS THE MOST EFFECTIVE WAY TO SURVIVE.

NOT ALL INDIVIDUALS ARE ALIKE: EACH ONE HAS A SLIGHTLY DIFFERENT COMBI-NATION OF GENETIC TRAITS, AND SOME OF THESE COMBINATIONS GIVE THEIR OWNERS A **SELECTIVE ADVANTAGE:** THE LUCKY ORGANISM IS BETTER ABLE TO GET FOOD, EVADE PREDATORS, WITHSTAND HEAT AND COLD, REPRODUCE, ETC.

AND WHICH ONE OF US HAS THE ADVANTAGE, COUSIN?

THE RESULT IS **DIFFERENTIAL REPRODUCTION:** THE FAVORED INDIVIDUALS BREED MORE OFFSPRING... SOME OF THEIR OFFSPRING INHERIT THE GOOD GENES, SO THEY BREED MORE PRODUCTIVELY TOO. AFTER SEVERAL GENERATIONS, THE BETTER-ADAPTED TYPE MAKES UP MOST OF THE POPULATION.

YOU DO!

SOMETIMES THE DESCENDANTS WILL
EVOLVE INTO AN ENTIRELY NEW SPECIES.

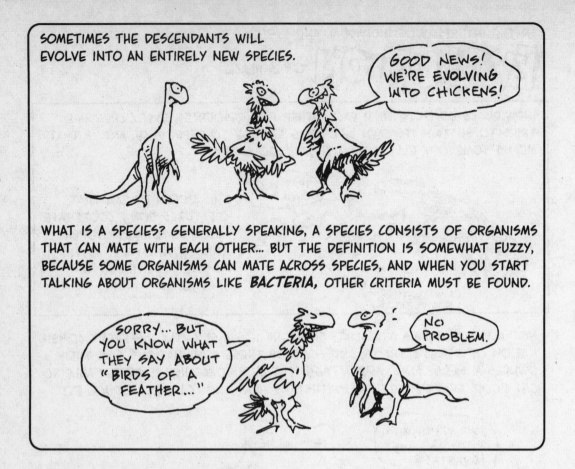

GOOD NEWS!
WE'RE EVOLVING
INTO CHICKENS!

WHAT IS A SPECIES? GENERALLY SPEAKING, A SPECIES CONSISTS OF ORGANISMS
THAT CAN MATE WITH EACH OTHER... BUT THE DEFINITION IS SOMEWHAT FUZZY,
BECAUSE SOME ORGANISMS CAN MATE ACROSS SPECIES, AND WHEN YOU START
TALKING ABOUT ORGANISMS LIKE *BACTERIA*, OTHER CRITERIA MUST BE FOUND.

SORRY... BUT
YOU KNOW WHAT
THEY SAY ABOUT
"BIRDS OF A
FEATHER..."

NO
PROBLEM.

EXACTLY HOW SPECIATION OCCURS
IS NOT WELL UNDERSTOOD. MOST
BIOLOGISTS BELIEVE IN

allopatric

("OTHER PLACE") SPECIATION: A
SMALL POPULATION BECOMES
GEOGRAPHICALLY ISOLATED IN
SOME WAY. BREEDING ONLY
AMONG THEMSELVES, ITS
MEMBERS EVOLVE AWAY FROM
THE ANCESTRAL TYPE. HUMANS
AND CHIMPANZEES DIVERGED, IT
IS BELIEVED, BECAUSE THE
ANCESTRAL SPECIES WAS DIVIDED
BY AFRICA'S GREAT RIFT VALLEY.

'ALLO,
PATRICK!

Sympatric

("SAME PLACE") SPECIATION CAN INVOLVE **SEASONAL** OR **HABITAT ISOLATION**— POTENTIAL MATES AREN'T IN THE SAME PLACE AT THE SAME TIME—OR **BEHAVIORAL ISOLATION**, FOR EXAMPLE, WHEN A COURTSHIP RITUAL DEVELOPS THAT APPEALS TO SOME BUT NOT TO ALL.

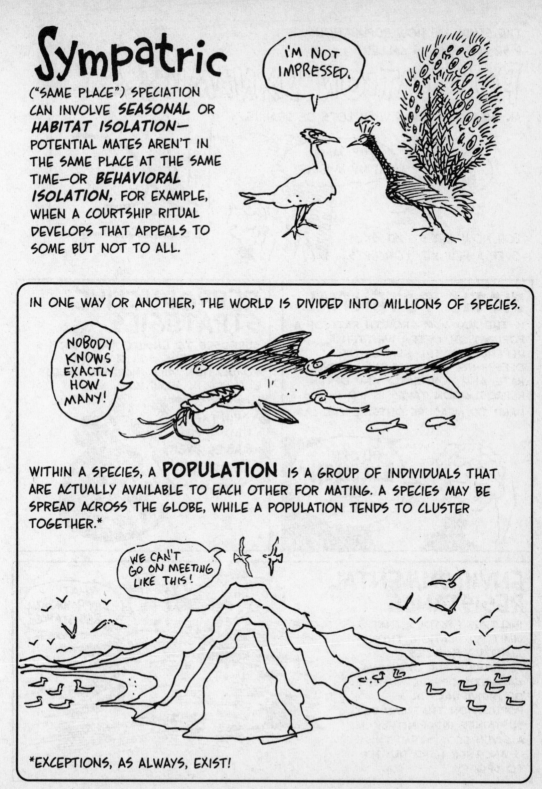

I'M NOT IMPRESSED.

IN ONE WAY OR ANOTHER, THE WORLD IS DIVIDED INTO MILLIONS OF SPECIES.

NOBODY KNOWS EXACTLY HOW MANY!

WITHIN A SPECIES, A **POPULATION** IS A GROUP OF INDIVIDUALS THAT ARE ACTUALLY AVAILABLE TO EACH OTHER FOR MATING. A SPECIES MAY BE SPREAD ACROSS THE GLOBE, WHILE A POPULATION TENDS TO CLUSTER TOGETHER.*

WE CAN'T GO ON MEETING LIKE THIS!

*EXCEPTIONS, AS ALWAYS, EXIST!

THE STUDY OF HOW POPULATIONS
RISE AND FALL IS CALLED

POPULATION DYNAMICS,

AN INEXACT SCIENCE WITH LOTS OF GRAPHS.

WOW! GREAT PICTURE!

YUP. MADE IT UP MYSELF.

FOR NOW, LET'S JUST SKIM
OVER A FEW KEY CONCEPTS.

BIOTIC POTENTIAL

IS THE MAXIMUM GROWTH RATE OF A
POPULATION, OFTEN WRITTEN r_{max}. IT
DEPENDS ON THE NUMBER OF
OFFSPRING, THEIR AVERAGE SURVIVAL
RATE, AND HOW EARLY AND OFTEN
REPRODUCTION TAKES PLACE. r_{max} IS
HARD TO MEASURE OUTSIDE THE LAB.

BUT AN **EXCELLENT** CONCEPT, REALLY!

REPRODUCTIVE STRATEGIES ARE

SUPPOSED TO ENSURE THAT BIRTHS
EXCEED DEATHS, SO THAT THE
POPULATION WILL INCREASE UNLESS
CHECKED IN SOME WAY.

WHAT IF I HAD 3000 BABIES EVERY 6 MONTHS...?

ENVIRONMENTAL RESISTANCE

INCLUDES FACTORS THAT
LIMIT POPULATION. THESE
LIMITING FACTORS
DETERMINE THE *CARRYING
CAPACITY,* OR THE NUMBER
OF INDIVIDUALS IN A
POPULATION THAT CAN BE
SUSTAINED INDEFINITELY IN
A GIVEN ECOSYSTEM. THIS
IS ANOTHER HARD NUMBER
TO SPECIFY.

COUGH

WHAT ENVIRONMENTAL RESISTANCE?

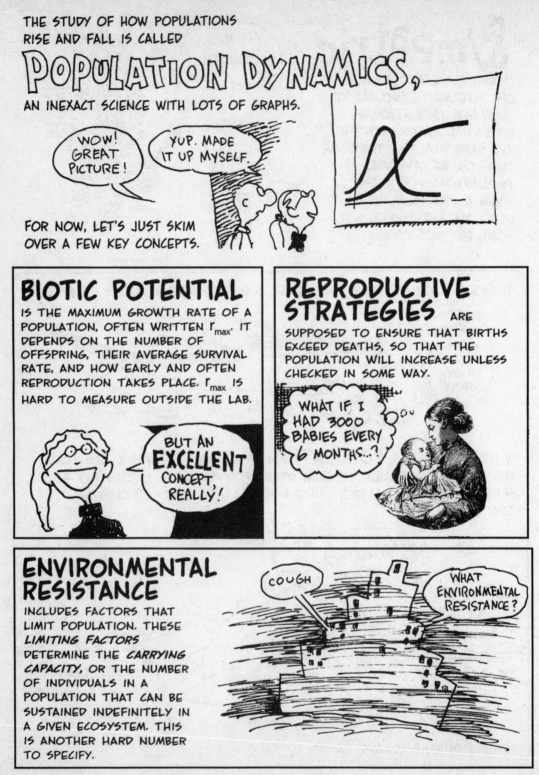

EXTRINSIC LIMITING FACTORS COME FROM OUTSIDE THE POPULATION. THEY CAN BE **ABIOTIC** (NONLIVING), SUCH AS AVAILABILITY OF CHEMICAL NUTRIENTS, THE LEVEL OF LIGHT, WATER, ETC. TOO MUCH OR TOO LITTLE OF ANY ONE ABIOTIC FACTOR CAN LIMIT POPULATION, EVEN IF ALL OTHER FACTORS ARE AT OPTIMAL LEVELS.

IT'S COOL IN HERE, BUT...

BIOTIC EXTRINSIC FACTORS INCLUDE FOOD SUPPLY, PREDATORS, DISEASE, AND THE GENERAL SUITABILITY OF THE ORGANIC ENVIRONMENT.

DON'T WORRY... IT'S ONLY A BIOTIC EXTRINSIC LIMITING FACTOR...

INTRINSIC LIMITING FACTORS COME FROM WITHIN THE POPULATION: **REPRODUCTIVE RATE, ADAPTABILITY,** AND EVEN **TERRITORIALITY:** GROUPS OR INDIVIDUALS MAY STAKE OUT A TERRITORY, FROM WHICH OTHERS IN THE SPECIES ARE EXCLUDED. THIS DETERMINES A POPULATION DENSITY AND HENCE A MAXIMUM LEVEL.

SOCIAL HIERARCHY CAN ALSO LIMIT POPULATION: IN MANY SPECIES, A DOMINANT MALE EJECTS LOWER-STATUS MALES FROM THE GROUP AND MAY EVEN KILL THEIR OFFSPRING.

BIG DADDY BULL APE HAS SOME STRONG OPINIONS!

DIFFERENT SPECIES ADOPT DIFFERENT REPRODUCTIVE STRATEGIES, DEPENDING ON WHAT LIMITING FACTORS APPLY. THE TWO MOST EXTREME STRATEGIES ARE:

r SELECTION: HAVE MASSES OF OFFSPRING IN ONE REPRODUCTIVE SHOT, AND THEN DIE. THIS STRATEGY WORKS WELL IN AN UNCROWDED ENVIRONMENT, WHERE POPULATION MAY EXPAND RAPIDLY, A HAZARDOUS ENVIRONMENT WHERE FEW SURVIVE, OR A RAPIDLY CHANGING ENVIRONMENT WHERE SWIFT ADAPTATION IS NEEDED.

GOOD LUCK, KIDS, AND GOOD·BYE!

K SELECTION: AN INDIVIDUAL REPRODUCES REPEATEDLY, HAS FEW OFFSPRING EACH TIME, CARES FOR THE YOUNG, AND HAS HIGH EXPECTATION FOR THEIR SURVIVAL. THIS STRATEGY OFTEN WORKS WELL WHEN POPULATION IS NEAR THE ENVIRONMENT'S CARRYING CAPACITY, BUT IT CAN BE RISKY. IF POPULATION DROPS TOO FAR, K SELECTION IS A SLOW WAY TO BUILD IT BACK UP.

* * * * * * * * *
ORANGUTANS, WITH JUST ONE OFFSPRING EVERY FIVE YEARS, ARE A PRIME EXAMPLE.
* * * * * * * * *

OR A PRIMATE EXAMPLE!

THEN THERE IS A RANGE OF STRATEGIES IN BETWEEN.

GIVEN THESE INFLUENCES ON POPULATION, WE CAN DRAW GRAPHS SHOWING PATTERNS OF POPULATION GROWTH.

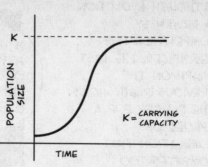

THERE MAY BE ENOUGH INTRINSIC AND EXTRINSIC CHECKS TO CAUSE THE POPULATION TO LEVEL OFF AT K.

WHEN A POPULATION IS SMALL COMPARED TO THE CARRYING CAPACITY, IT GROWS MORE OR LESS RAPIDLY, DEPENDING ON ITS INTRINSIC RATE OF REPRODUCTION. WHAT HAPPENS NEXT DEPENDS ON MANY FACTORS.

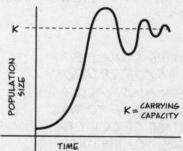

MORE TYPICALLY, POPULATION OVERSHOOTS K, IN WHICH CASE IT MAY WOBBLE BACK TO AN EQUILIBRIUM LEVEL...

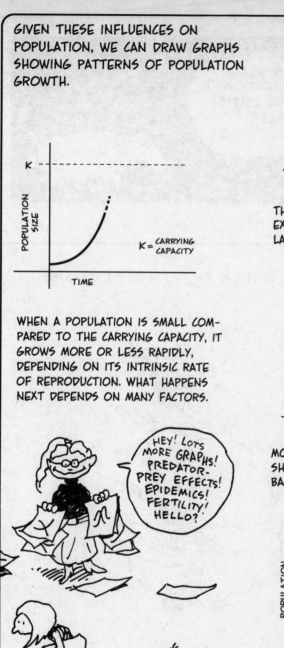

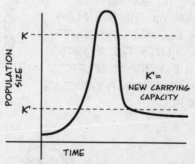

...OR HAVE A CATASTROPHE: POPULATION GROWTH DISRUPTS THE ECOSYSTEM SO BADLY THAT ITS CARRYING CAPACITY IS REDUCED, AND THE SPECIES SUFFERS A DIE-OFF FROM WHICH IT NEVER FULLY RECOVERS.

ALTHOUGH EVOLUTION IS DRIVEN BY COMPETITION FOR RESOURCES, THE LAST GRAPH ON THE PREVIOUS PAGE SHOWS THE DANGER FOR A SPECIES OR POPULATION THAT COMPETES TOO "SUCCESSFULLY" IN THE GAME OF LIFE.

THIS WAS A PRETTY NICE PLACE UNTIL WE ATE EVERYTHING!

IN THE WEB OF LIFE, EACH SPECIES DEPENDS ON THE SURVIVAL OF MANY OTHERS.

A STRIKING EXAMPLE IS GIVEN BY *COEVOLUTION*: TWO (OR MORE) SPECIES MAY *COEVOLVE* TO TAKE ADVANTAGE OF EACH OTHER. FLOWERS, FOR INSTANCE, HAVE STRUCTURES THAT RUB OFF POLLEN ON NECTAR-SEEKING BEES, WHILE BEE HIND LEGS ARE PERFECTLY SUITED TO ACCUMULATE POLLEN. AS THE BEE FLITS FROM FLOWER TO FLOWER, IT FERTILIZES THE PLANTS, WHICH DEPEND COMPLETELY ON BEES FOR REPRODUCTIVE SUCCESS.

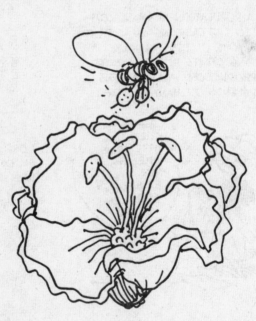

IN OTHER WORDS, DIVERSITY IS HEALTHY!

BIODIVERSITY COMES IN THREE FLAVORS:

GENETIC DIVERSITY

IS DIVERSITY WITHIN A SPECIES. DIFFERENT INDIVIDUALS DO HAVE SMALL DIFFERENCES IN THEIR GENETIC MAKEUP. GENETIC DIVERSITY IS HEALTHY BECAUSE IT MAKES A SPECIES MORE ADAPTABLE TO ENVIRONMENTAL CHANGE.

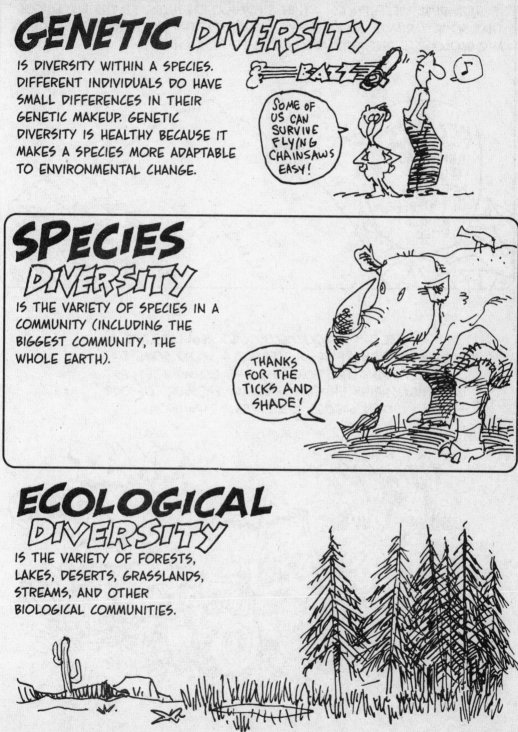

BAZZ

SOME OF US CAN SURVIVE FLYING CHAINSAWS EASY!

SPECIES DIVERSITY

IS THE VARIETY OF SPECIES IN A COMMUNITY (INCLUDING THE BIGGEST COMMUNITY, THE WHOLE EARTH).

THANKS FOR THE TICKS AND SHADE!

ECOLOGICAL DIVERSITY

IS THE VARIETY OF FORESTS, LAKES, DESERTS, GRASSLANDS, STREAMS, AND OTHER BIOLOGICAL COMMUNITIES.

EVOLUTION NOT ONLY GENERATES GENETIC DIVERSITY BUT ALSO DEPENDS ON IT. INDIVIDUAL DIFFERENCES WITHIN A POPULATION INCREASE THE LIKELIHOOD THAT SOME VARIANTS WILL SURVIVE CHANGES IN THE ENVIRONMENT. SPECIES AND ECOLOGICAL DIVERSITY ALSO GIVE LIFE RESILIENCY.

SINCE THESE LOCAL ECOLOGIES REGULATE THE CIRCULATION OF RESOURCES, LET'S SPEND SOME TIME REVIEWING THE WORLD'S LIVING COMMUNITIES, AS THEY SHRINK UNDER THE STEADY PRESSURE OF OUR OWN SPECIES' K-STRATEGIC EXPANSION.

· CHAPTER 4 ·

COMMUNITIES WET...

LET'S START IN THE OCEAN, WHERE LIFE BEGAN.

HMMM... MAYBE I OUGHT TO GO **BACK**....

THE OCEAN COVERS OVER **70%** OF THE EARTH AND IS CONSTANTLY FLOWING. OCEAN CURRENTS MOVE MASSES OF WATER LONG DISTANCES: THE WARM GULF STREAM ALONE CARRIES **50 TIMES MORE WATER** THAN ALL THE WORLD'S RIVERS COMBINED. SINCE WATER HOLDS HEAT, OCEAN CURRENTS DISTRIBUTE SOLAR ENERGY, WHILE ITS SHEER BULK STABILIZES CLIMATIC CHANGE. THE OCEAN ALSO SERVES AS A RESERVOIR OF DISSOLVED GASES THAT REGULATES THE COMPOSITION OF THE ATMOSPHERE.

MORE THAN JUST A FISH TANK!

OF THE 1.4 MILLION SPECIES THAT HAVE BEEN CLASSIFIED TO DATE, ONLY 250,000, OR 35%, LIVE IN THE OCEAN. THIS IS NOT BECAUSE THE OCEANS ARE LESS DIVERSE, BUT BECAUSE WE HAVE SPENT LESS TIME THERE LOOKING FOR NEW SPECIES. DESPITE OUR IGNORANCE, THERE ARE A FEW THINGS WE CAN SAY ABOUT UNDERWATER COMMUNITIES.

FOR THE MOST PART, AQUATIC LIFE PRACTICES r SELECTION: ADULTS BREED PROFUSELY, AND THE YOUNG RARELY REACH MATURITY. THIS IS THE HABIT OF MOST FISH AND EVERYTHING LOWER DOWN THE EVOLUTIONARY SCALE. MAMMALS LIKE SEALS, DOLPHINS, AND WHALES ARE THE EXCEPTIONAL K-BREEDERS.

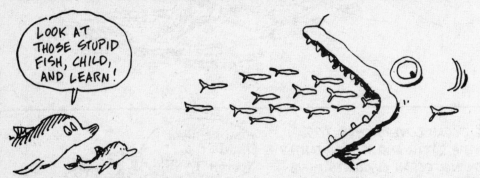

COMPARED WITH LANDLOCKED ECOSYSTEMS, THE OCEANS HAVE MANY LAYERS OF CARNIVORY: PREDATORS PREYING ON PREDATORS PREYING ON PREDATORS. SOME SCIENTISTS BELIEVE THAT THE OCEANS WERE MAXIMALLY HARVESTED FROM WITHIN BEFORE HUMANS STARTED FISHING.

IN THE OCEAN, AS EVERYWHERE, ALL LIFE DEPENDS ON PLANTS, THE PRODUCERS. WE USE THE TERM **NET PRIMARY PRODUCTION** TO DESCRIBE THE TOTAL FOOD MADE AVAILABLE BY PLANTS.

IN THE OCEAN, THIS INCLUDES SEAWEED, MICROSCOPIC ALGAE, AND BACTERIA!

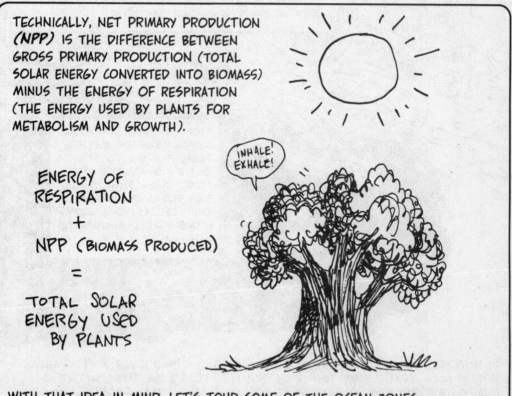

TECHNICALLY, NET PRIMARY PRODUCTION **(NPP)** IS THE DIFFERENCE BETWEEN GROSS PRIMARY PRODUCTION (TOTAL SOLAR ENERGY CONVERTED INTO BIOMASS) MINUS THE ENERGY OF RESPIRATION (THE ENERGY USED BY PLANTS FOR METABOLISM AND GROWTH).

INHALE! EXHALE!

ENERGY OF RESPIRATION
+
NPP (BIOMASS PRODUCED)
=
TOTAL SOLAR ENERGY USED BY PLANTS

WITH THAT IDEA IN MIND, LET'S TOUR SOME OF THE OCEAN ZONES...

OPEN OCEAN

THE OCEAN FOOD WEB DEPENDS ENTIRELY ON PLANTS IN THE SUNLIT UPPER LAYER (THE *EUPHOTIC ZONE*).

THE VAST BULK OF THESE PLANTS ARE DRIFTING, MICROSCOPIC *PHYTOPLANKTON,* WHICH NOURISH SMALL ANIMALS CALLED *ZOOPLANKTON,* WHICH IN TURN FEED SMALL FISH AND HUGE BALEEN WHALES. THE SMALL FRY ARE EATEN BY LARGER FISH, SEALS, PORPOISES, BIRDS, ETC. DEAD ORGANISMS FALL TO THE SEA FLOOR TO NOURISH SCAVENGERS LIKE CRABS AND URCHINS AND DECOMPOSERS LIKE BACTERIA.

THE TWILIGHT *BATHYAL ZONE,* 200-500 METERS DEEP, IS HOME TO SQUID, OCTUPI, SHRIMP, AND A FEW TOUGH FISH, WHILE THE TOTAL DARKNESS OF THE DEEP *ABYSSAL ZONE* IS NOT WELL KNOWN TO HUMANS.

DESPITE ALL THIS ACTION, OCEAN LIFE IS SPREAD THINLY OVER A VAST AREA. NPP IN AN AVERAGE PART THE OPEN OCEAN IS VERY LOW, LIKE THAT OF A DESERT. BUT BECAUSE THE OCEAN IS SO LARGE, THE TOTAL PRODUCTION OF FOOD IS IMMENSE.

MOVING IN TOWARD SHORE, WE
ENTER A MUCH MORE PRODUCTIVE
REGION, WHERE LIFE IS DENSE AND
ALL IS SUNNY: THE

COASTAL ZONE.

IT EXTENDS FROM THE OCEAN'S
HIGH-TIDE LINE OUT TO THE EDGE OF
THE CONTINENTAL SHELF, ABOUT 10%
OF THE OCEAN'S AREA.

MANY COASTAL ZONES HAVE *CORAL REEFS,* THE OCEAN'S MOST DIVERSE ECO-
SYSTEMS. A SINGLE REEF MAY HOUSE 3000 CORAL SPECIES AND COUNTLESS
KINDS OF FISH, WORMS, SEAWEED, AND OTHER LIFE FORMS. THE *PRIMARY
PRODUCERS* HERE ARE ONE-CELLED, GREEN *DINOFLAGELLATES* THAT TAKE
SHELTER AMONG THE CORAL, USING MINERAL NUTRIENTS SHED BY THEIR HOSTS.
THE CORAL RESPONDS BY EATING ORGANIC CARBON COMPOUNDS EXCRETED BY
THEIR GUESTS. THE REEF ITSELF IS BUILT UP OF LIMESTONE SHELLS FORMED
BY THE LIVING CORAL, AND WHEN THE CORAL DIES, THE SHELLS REMAIN. LIME-
STONE (CALCIUM CARBONATE) SEQUESTERS CARBON FROM THE ATMOSPHERE,
BUFFERING CLIMATE CHANGE. DESPITE THEIR DIVERSITY, CORAL REEFS ARE SLOW-
GROWING AND FRAGILE, AND MANY ARE FALLING VICTIM TO HUMAN ACTIVITIES.

MOVING ASHORE, **COASTAL WETLANDS** ARE A MIX OF BAYS, SALT MARSHES, AND MUDFLATS, WHERE GRASSES DOMINATE THE FOOD PRODUCTION.

IN WARM CLIMATES, **MANGROVE SWAMPS**—A PARTLY SUBMERGED COLLECTION OF ABOUT 50 SPECIES OF SALT-TOLERANT TREES AND SHRUBS— ARE THE WORLD'S MOST PRODUCTIVE ECOSYSTEM.

THE MANGROVE'S TALL ROOTS FUNNEL OXYGEN INTO THE MUCK BELOW, AND MANY SEAGOING SPECIES SPEND THEIR YOUTH IN A MANGROVE SWAMP: THERE IS PLENTY TO EAT, AND IT'S EASY TO HIDE IN THE VEGETATION.

STREAMS

FLOW DOWN FROM THE HILLS WHERE THE WATER, ORIGINATING IN THE OCEAN, FALLS AFTER EVAPORATING.

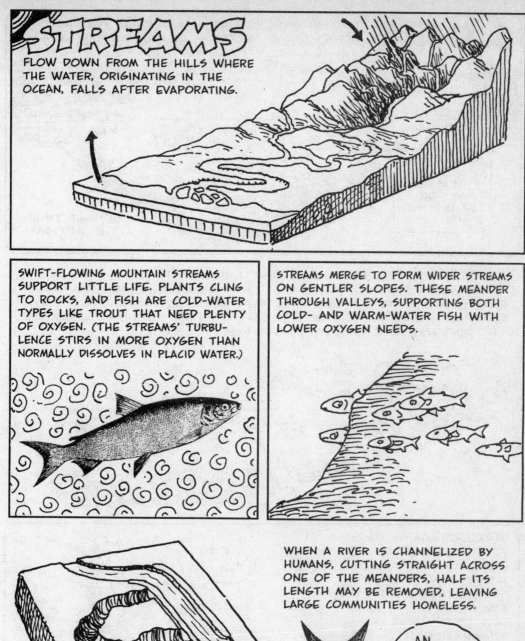

SWIFT-FLOWING MOUNTAIN STREAMS SUPPORT LITTLE LIFE. PLANTS CLING TO ROCKS, AND FISH ARE COLD-WATER TYPES LIKE TROUT THAT NEED PLENTY OF OXYGEN. (THE STREAMS' TURBULENCE STIRS IN MORE OXYGEN THAN NORMALLY DISSOLVES IN PLACID WATER.)

STREAMS MERGE TO FORM WIDER STREAMS ON GENTLER SLOPES. THESE MEANDER THROUGH VALLEYS, SUPPORTING BOTH COLD- AND WARM-WATER FISH WITH LOWER OXYGEN NEEDS.

WHEN A RIVER IS CHANNELIZED BY HUMANS, CUTTING STRAIGHT ACROSS ONE OF THE MEANDERS, HALF ITS LENGTH MAY BE REMOVED, LEAVING LARGE COMMUNITIES HOMELESS.

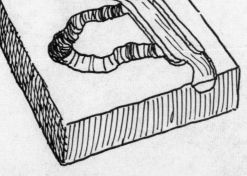

AN ENGINEERING MARVEL!

LAKES

FORM WHEN RUNOFF OR GROUNDWATER FILLS A DEPRESSION IN THE EARTH. A LAKE HAS FOUR ZONES:

LITTORAL ZONE: NEAR THE SHORE

LIMNETIC ZONE: OPEN WATER WITH ENOUGH LIGHT FOR PLANTS TO GROW.

PROFUNDAL ZONE: COLD AND DEEP

BENTHIC ZONE: THE BOTTOM

A **EUTROPHIC** ("TRUE FOOD") LAKE IS WELL-NOURISHED, OFTEN SHALLOW, AND MURKY WITH PLANKTON. VARIOUS FISH, SUCH AS BASS, PERCH, SUNFISH, AND PIKE, FLOURISH THERE. THE LITTORAL ZONE IS BROAD, A GOOD HABITAT FOR FROGS AND FISH. OXYGEN IS A COMMON LIMITING FACTOR FOR EUTROPHIC LAKES, AND THE PROFUNDAL ZONE IS OFTEN OXYGEN-FREE.

IN **OLIGOTROPHIC** ("FEW FOOD") LAKES, THERE IS USUALLY A SHORTAGE OF NITRATE AND PHOSPHATE. THE WATER IS CRYSTALLINE AND OFTEN DEEP. AN OLIGOTROPHIC LAKE CAN BE RECOGNIZED BY ITS NARROW SHORE AND CLEAR WATER.

INLAND WETLANDS,

LIKE COASTAL WETLANDS, ARE HIGHLY PRODUCTIVE. THESE INCLUDE BOGS, MARSHES, PRAIRIE POTHOLES, FLOODPLAINS, AND WET ARCTIC TUNDRA IN THE SUMMER. WATER IS STILL AND WARM; NUTRIENTS ARE TRAPPED BY PLANT STEMS; PLANKTON ABOUND. MAMMALS, BIRDS, AMPHIBIANS, AND INSECTS ALL FEAST IN WETLANDS.

WETLANDS REGULATE STREAM FLOW BY ACTING LIKE A SPONGE: IN THE DRY SEASON, THEY MAY APPEAR PARCHED AND CRACKED, BUT THEY REMAIN POROUS.

DURING RAIN OR SPRING MELT, THEY SWELL UP, STORING WATER AND RELEASING IT SLOWLY DOWNSTREAM OR ALLOWING IT TO PERCOLATE DOWN INTO THE GROUNDWATER.

IN THE ABSENCE OF WETLANDS, FLOODING AND EROSION INCREASE. UNFORTUNATELY, ABOUT HALF THE WETLANDS IN THE U.S. HAVE BEEN DRAINED OR FILLED FOR AGRICULTURE.

ALTHOUGH LIFE BEGAN IN THE WATER, NOT ALL LIVING COMMUNITIES ARE WATERY. SEVERAL HUNDRED MILLION YEARS AGO, LIFE INVADED THE LAND. DESPITE EVERY-THING YOU HEAR ABOUT LUNGFISH AT THE SCIENCE MUSEUM, THEY COULD NOT POSSIBLY HAVE BEEN THE FIRST LIFE ASHORE—BECAUSE THEY ARE CONSUMERS!

THE FIRST LAND-BASED LIFE MUST HAVE BEEN PRODUCERS, PROBABLY ONE-CELLED PLANTS AND SOME BACTERIA. ONLY AFTER THERE WAS ENOUGH GREEN SCUM ON THE ROCKS COULD SMALL ANIMALS—ALSO ONE-CELLED—LEAD THE CHARGE OF THE CONSUMERS.

SINCE THEN, THESE EARLY ALGAE, LICHEN, MOSSES, AND FERNS, ALONG WITH THEIR CONSUMERS, THE INSECTS, ARTHROPODS, AND FISH, HAVE EVOLVED INTO MILLIONS OF DIFFERENT SPECIES, WHOSE COMMUNITIES ARE THE SUBJECT OF OUR NEXT CHAPTER...

❖ CHAPTER 5 ❖

...AND DRY

ON LAND, BASIC LIFE COMMUNITIES ARE CALLED **BIOMES.** BIOMES
DEPEND ON CLIMATE AND GEOGRAPHY: WHERE CLIMATE IS DIFFERENT, SO
ARE THE PLANTS AND ANIMALS, WHILE SIMILAR BIOMES MAY EXIST IN WIDELY
SEPARATED AREAS THAT HAVE SIMILAR ENVIRONMENTAL CONDITIONS.

AS A GENERAL RULE, CLIMATE COOLS AS YOU HEAD AWAY FROM THE EQUATOR
AND TOWARD THE POLES... AND ALSO AS YOU RISE ABOVE SEA LEVEL. RAINFALL
IS HEAVY AT THE EQUATOR, LIGHTER AT THE TROPICS OF CANCER AND CAPRICORN,
AND RISES AGAIN POLEWARDS.

BIOMES DEPEND ON LATITUDE AND ALTITUDE!

PRECIPITATION DETERMINES WHETHER THE LAND IS DESERT, GRASSLAND, OR
FOREST, AND TEMPERATURE ALSO INFLUENCES THE STRATEGIES OF BOTH PLANTS
AND ANIMALS. WE CAN SUMMARIZE THIS BY LUMPING ALL THE WORLD'S BIOMES
INTO A SUPERCONTINENT, SHOWING WHAT EVOLUTION AND GEOGRAPHY HAVE
WROUGHT ON A GRAND SCALE.

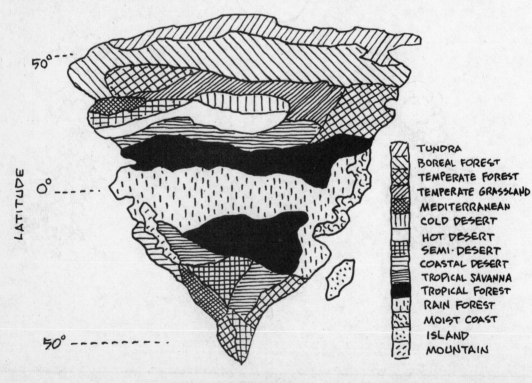

TUNDRA
BOREAL FOREST
TEMPERATE FOREST
TEMPERATE GRASSLAND
MEDITERRANEAN
COLD DESERT
HOT DESERT
SEMI-DESERT
COASTAL DESERT
TROPICAL SAVANNA
TROPICAL FOREST
RAIN FOREST
MOIST COAST
ISLAND
MOUNTAIN

HERE IS A TOUR OF THE WORLD'S MAJOR BIOMES:

POLAR GRASSLAND
(OR ARCTIC TUNDRA)

IS AN ICY, TREELESS BIOME NORTH OF THE TREELINE AND BELOW THE ARCTIC CIRCLE (AND JUST ABOVE THE TIMBERLINE ON MOUNTAINS). THE TUNDRA'S LIMITING FACTORS ARE **HEAT** AND **LIGHT** (OR RATHER, COLD AND DARKNESS!), SO FEW PRODUCER SPECIES CAN LIVE HERE. THE SOIL IS PERMANENTLY FROZEN, ICING OUT TREE ROOTS. WINTERS ARE FIERCE.

DURING THE BRIEF SUMMER THAW, SURFACE SOIL BECOMES A QUAGMIRE OF PUDDLES AND BOGS; LOW-GROWING PLANTS BUD AND BLOSSOM. HORDES OF INSECTS HATCH; BIRDS ARRIVE TO EAT THE BUGS. THE PERMANENT RESIDENTS ARE MOSTLY BURROWING PLANT-EATERS THAT HUDDLE UNDER THE SNOW IN WINTER. BIG HERBIVORES VISIT IN THE SUMMER AND THEN HEAD SOUTH. FOX, LYNX, AND BEAR LIVE HERE TOO.

THE SLOW-GROWING PLANTS AND SHALLOW SOIL MAKE TUNDRA THE MOST FRAGILE BIOME: 100-YEAR-OLD WAGON TRACKS ARE STILL VISIBLE TODAY.

BOREAL FOREST
(OR "TAIGA")

COLD AND WET, THE TAIGA STRETCHES ACROSS NORTH AMERICA AND EURASIA, COVERING 11% OF ALL LAND. WINTERS ARE LONG AND COLD, BUT SUMMERS ARE LONGER AND WARMER THAN IN THE TUNDRA. BOREAL FORESTS INCLUDE YOSEMITE, SEQUOIA, AND YELLOWSTONE NATIONAL PARKS.

DOMINATED BY CONIFEROUS (CONE-BEARING) EVERGREEN TREES, THE RELATIVELY INFERTILE FOREST FLOOR IS STREWN WITH NEEDLES AND LEAF LITTER, INTERRUPTED BY A FEW HARDY SHRUBS AND PLANTS. LARGE HERBIVORES INCLUDE MOOSE, MULE DEER, CARIBOU, AND ELK; SMALL ONES INCLUDE HARE, SQUIRRELS, AND OTHER RODENTS. INSECT DIVERSITY IS LOW, WITH BUTTERFLIES, BEETLES, WASPS, AND FLIES. MAJOR PREDATORS INCLUDE THE TIMBERWOLF, LYNX, FOX, MARTEN, WOLVERINE, MINK, OTTER, ERMINE, AND WEASEL.

TAIGA IS THREATENED BY HEAVY LOGGING, SINCE CONIFEROUS TREES GROW RAPIDLY, GIVING THEM HIGH ECONOMIC VALUE FOR PULP AND LUMBER PRODUCTS.

TEMPERATE DECIDUOUS FOREST

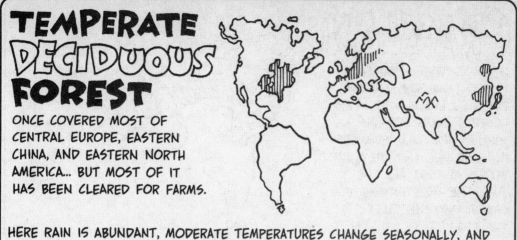

ONCE COVERED MOST OF CENTRAL EUROPE, EASTERN CHINA, AND EASTERN NORTH AMERICA... BUT MOST OF IT HAS BEEN CLEARED FOR FARMS.

HERE RAIN IS ABUNDANT, MODERATE TEMPERATURES CHANGE SEASONALLY, AND THE GROWING SEASON IS 4-6 MONTHS.

THE DECIDUOUS FOREST IS STRATIFIED, OR LAYERED, WITH A HIGH CANOPY, AN UNDERSTORY OF SMALLER HARDWOODS, A SHRUB LAYER, THEN GROUND-COVERING HERBS, AND A FINAL LAYER OF MOSS AND LICHEN. AS TEMPERATURES DROP IN THE FALL, SO DO THE LEAVES, DECOMPOSING TO PRODUCE RICH SOIL THAT SUPPORTS A WIDE VARIETY OF PLANTS.

WARBLE HOOT TWITTER FLAP

ROT SIMMER DECAY FERMENT PERCOLATE

DIFFERENT LAYERS PROVIDE MORE HABITAT FOR HERBIVORES (WHITETAIL DEER, BLACK BEAR, BEAVER, PORCUPINE, OPOSSUM, RACCOON, SKUNK, CHIPMUNK, GRAY SQUIRREL, SHREW, AND THE COTTONTAIL RABBIT), SO THERE ARE MORE CARNIVORES (WOLVES, BOBCATS, GRAY FOXES, MOUNTAIN LIONS). THIS BIOME HAS BEEN THE MOST REDUCED BY HUMAN ACTIVITY, AND RELATIVELY FEW VIRGIN STANDS REMAIN.

GRASSLAND

IS FOUND WHERE RAINFALL IS TOO LOW FOR FORESTS BUT HIGH ENOUGH TO KEEP DESERTS FROM FORMING. WINTERS ARE COLD, SUMMERS HOT AND DRY, AND THE WIND BLOWS ALMOST ALL THE TIME. THE MOISTER THE GRASSLAND, THE TALLER THE GRASS.

BEFORE AGRICULTURE, THE AMERICAN GREAT PLAINS WERE DOMINATED BY LARGE HERBIVORES: BISON, ANTELOPE, ELK; SMALL HERBIVORES, LIKE PRAIRIE DOGS AND JACKRABBIT; AND PREDATORS, INCLUDING WOLF, COYOTE, PANTHER, AND PEOPLE.

TRY NOT TO DRAW ATTENTION TO YOURSELF THAT WAY!

BUT THIS TALLGRASS SOIL IS AMONG THE WORLD'S RICHEST, AND NOW NEARLY EVERY ACRE OF IT IS PLANTED IN WHEAT AND CORN.

WELL, I'M A LARGE HERBIVORE, TOO!

HAVE I SEEN YOU HERE-BEVORE?

GRASSLANDS ARE UNDER CONSTANT STRESS FROM WIND, SUN, AND TEMPERATURE EXTREMES. DROUGHT OR MISUSE CAN LEAD TO WIND EROSION, STRIPPING IRREPLACEABLE TOPSOIL AND TURNING GRASSLAND INTO BARREN DESERT.

DESERTS

COVER ALMOST 1/4 OF THE LAND'S SURFACE. **COLD DESERTS,** LIKE THE GOBI, HAVE WARM SUMMERS AND FRIGID WINTERS. **TEMPERATE DESERTS,** LIKE THE MOJAVE IN SOUTHERN CALIFORNIA, HAVE HOT SUMMERS AND COOL WINTERS. **TROPICAL DESERTS,** LIKE THE SOUTHERN SAHARA, ARE JUST PLAIN HOT AND SUPPORT LITTLE LIFE. ALL DESERTS HAVE LITTLE WATER AND LOW SPECIES DIVERSITY. DESERT PLANTS, WHICH HAVE EVOLVED TO COLLECT AND CONSERVE PRECIOUS WATER, GROW SLOWLY, SO DESERT ECOSYSTEMS ARE FRAGILE.

TROPICAL RAINFOREST,

THE MOST DIVERSE OF
ALL BIOMES, COVERS
ABOUT 15% OF THE LAND
AREA—AND FALLING. IT
GROWS NEAR THE
EQUATOR, WHERE RAIN
AND HEAT ARE PLENTIFUL.

THE LIMITING FACTOR HERE IS
SUNLIGHT, BECAUSE THE DENSE LEAF
CANOPY SHADES THE FOREST BENEATH.
THE AIR IS STILL, PREVENTING PLANTS
FROM USING WIND POLLINATION.
INSTEAD, TROPICAL FLOWERS HAVE
COEVOLVED WITH BIRDS, INSECTS, AND
BATS AS POLLINATORS.

THE GROUND IS NEARLY PLANT-
FREE, EXCEPT AT THE EDGES OF
CLEARINGS. ARMIES OF INSECTS
SPEEDILY CONSUME ANYTHING THAT
FALLS TO THE FOREST FLOOR.

MOST OF THE PLANT NUTRIENTS ARE ABOVE GROUND IN THE VEGETATION, NOT IN THE SOIL. *EPIPHYTES* (AIR-GROWING PLANTS, LIKE ORCHIDS) CLING TO THE TREES AND VINES, THEIR LEAVES TRAPPING POOLS OF RAINWATER, WHICH CONTAIN TINY ECOSYSTEMS OF THEIR OWN. INSECTS, BIRDS, FROGS, REPTILES, AND MAMMALS SWARM IN THE BRANCHES. *ONE RAINFOREST TREE* CAN HOLD MORE SPECIES THAN AN ENTIRE TAIGA!

IN PARTS OF THIS BIOME, SOIL IS ACTUALLY POOR, CONTAINING *LATERITES*, COMPOUNDS OF ALUMINUM AND IRON THAT BAKE INTO CONCRETE-LIKE SLABS WHEN EXPOSED TO THE SUN BY DEFORESTATION.

TROPICAL GRASSLAND,

OR **SAVANNA**, COVERS 11% OF THE LAND. IT OCCURS IN AREAS WITH TWO LONG DRY SEASONS, NO WINTER, AND ABUNDANT RAIN THE REST OF THE YEAR.

THE SAVANNA HAS BROAD EXPANSES OF TALL GRASSES AND SCATTERED TREES. TROPICAL SAVANNAS GROW QUICKLY, CREATING AS MUCH PLANT MATTER AS A TROPICAL RAINFOREST, FOOD FOR HUGE HERDS OF HERBIVORES.

BETWEEN ANY TWO BIOMES LIES A TRANSITION ZONE CALLED AN **ECOTONE.** THIS REFERS TO ANY REGION WHERE TWO BIOMES OR ECOSYSTEMS MEET. AN ECOTONE CONTAINS SPECIES FROM BOTH COMMUNITIES AND MAY HAVE SOME OTHER CHARACTERISTIC SPECIES ALL ITS OWN.

THE INCREASED VARIETY AND DENSITY OF SPECIES IN AN ECOTONE IS KNOWN AS THE **EDGE EFFECT.**

AND JUST TO MAKE LIFE EVEN MORE INTERESTING, THE EDGES HAVE A WAY OF MOVING AROUND: NOTHING STAYS PUT IN NATURE!

ALTHOUGH WE'VE DESCRIBED OUR BIOMES AS IF THEY WERE STABLE, IN FACT, THEY'RE ALWAYS CHANGING. SPECIES COME AND GO... THE ENVIRONMENT CHANGES... BORDERS SHIFT... NEW OPPORTUNITIES AND CHALLENGES ARISE...

IN PARTICULAR, LIFE CAN COLONIZE VIRGIN TERRITORY AND TURN BARREN LAND INTO A THRIVING ECOSYSTEM, A PROCESS KNOWN AS

ECOLOGICAL SUCCESSION

(OR COMMUNITY DEVELOPMENT).

PRIMARY SUCCESSION

OCCURS WHEN LIFE INVADES AN AREA WITH NO TRUE SOIL. FOR EXAMPLE:

COOLED LAVA

SUBSOIL EXPOSED BY A MUDSLIDE

A NEWLY DEPOSITED SANDBAR

A STRIP MINE

WE CAN EAT THAT!

AT FIRST, A FEW HARDY MICROBES, MOSSES, AND LICHENS INVADE THE AREA.

THESE PIONEER SPECIES ARE MOSTLY SMALL PRODUCERS, PLUS A FEW DECOMPOSERS. PIONEER PLANTS ARE SMALL, LOW-GROWING ANNUALS, **r**-STRATEGISTS THAT SPEND THEIR ENERGY MAKING LOTS OF SEEDS RATHER THAN DEVELOPING MUCH IN THE WAY OF ROOTS, STEMS, OR LEAVES.

EVENTUALLY, SOIL IS CREATED, WEEDS SPROUT, AND ANIMALS ARRIVE, SEEKING FOOD OR COVER.

SECONDARY SUCCESSION

BEGINS WHERE VEGETATION HAS BEEN REMOVED BUT SOIL IS INTACT.
FOR EXAMPLE:

ABANDONED FARMLAND

BURNED OR LOGGED FOREST

LAND RECLAIMED FROM FLOOD.

HERE NEW PLANTS CAN SPROUT WITHIN WEEKS. FIRST COME THE ANNUAL WEEDS...

SPRONG!!   SPROUT   SPRING

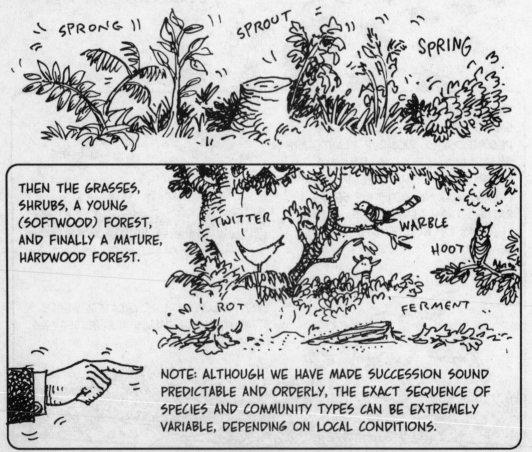

THEN THE GRASSES, SHRUBS, A YOUNG (SOFTWOOD) FOREST, AND FINALLY A MATURE, HARDWOOD FOREST.

TWITTER   WARBLE   HOOT   ROT   FERMENT

NOTE: ALTHOUGH WE HAVE MADE SUCCESSION SOUND PREDICTABLE AND ORDERLY, THE EXACT SEQUENCE OF SPECIES AND COMMUNITY TYPES CAN BE EXTREMELY VARIABLE, DEPENDING ON LOCAL CONDITIONS.

BY THE TIME AN ECOSYSTEM MATURES, IT HAS DIVERSE SPECIES, STABLE POPU-
LATIONS, AND COMPLEX INTERDEPENDENCIES AMONG ITS PLANTS AND ANIMALS.
TO SURVIVE, THE RESIDENTS NEED LARGE AREAS OF UNDISTURBED WILDERNESS.

ECOLOGISTS USED TO
BELIEVE THAT HIGHER
SPECIES DIVERSITY MEANT A
MORE STABLE ECOSYSTEM.
THE MORE SPECIES THERE
ARE, IT SEEMED, THE MORE
OPTIONS THE ECOSYSTEM
WOULD HAVE FOR RESPONDING
TO STRESS.

BUT AS IT HAPPENS, THE RELATIONSHIP BETWEEN SPECIES DIVERSITY AND STABILITY
CAN BE TENUOUS, EVEN NONEXISTENT. COMPLEX ECOSYSTEMS CAN BE FRAGILE!!

61

THIS IS BECAUSE ECOLOGICAL STABILITY ITSELF IS A COMPLEX PHENOMENON, WITH AT LEAST THREE DIFFERENT ASPECTS:

INERTIA, CONSTANCY, AND RESILIENCE.

INERTIA IS THE ABILITY OF AN ECO-SYSTEM TO RESIST CHANGE. A RAIN FOREST, FOR EXAMPLE, HAS HIGH INERTIA.

OOPS. HERE COME THE FIRE ANTS!

CONSTANCY IS THE ABILITY OF A LIVING SYSTEM, LIKE A POPULATION, TO PRESERVE ITS NUMBERS. THE HUMAN RACE HAS TERRIFIC CONSTANCY.

PLAGUE, FAMINE, OR WAR, WE JUST KEEP GOING!

RESILIENCE IS A SYSTEM'S ABILITY TO RESTORE ITSELF AFTER SUFFERING AN OUTSIDE DISTURBANCE. GRASSLANDS, FOR EXAMPLE, SPRING BACK AFTER A FIRE, BECAUSE SO MUCH OF THE PLANT MATTER IS SUBSURFACE ROOTS.

BUT THE DIVERSE, HIGH-INERTIA RAIN FOREST HAS VERY LOW RESILIENCE: ONCE CLEARED, THE FOREST IS GONE, BECAUSE SOIL NUTRIENTS AND THE WATER CYCLE CAN NO LONGER SUPPORT IT.

BY CONTRAST, GRASSLAND HAS LOW SPECIES DIVERSITY AND LOW INERTIA, BUT HIGH RESILIENCE.

NO PROBLEM.

ANOTHER MISCONCEPTION IS THAT ECOSYSTEMS HAVE A FAIRLY STABLE *EQUILIBRIUM* STATE, NEAR WHICH THEY REMAIN.

IN FACT, ECOSYSTEMS ARE RARELY, IF EVER, AT EQUILIBRIUM. NATURE CONSTANTLY FLUCTUATES.

CHANGE AND TURMOIL, RATHER THAN BALANCE, IS THE RULE.

CHOMP WHISK

POPULATIONS AND COMMUNITIES TEND TO SWING BETWEEN LIMITS BUT RARELY REMAIN CONSTANT.

IF DISTURBED, SYSTEMS MAY CHANGE AND OPERATE WITHIN *NEW LIMITS*...

AND NEVER RETURN TO SOME IMAGINED IDEAL EQUILIBRIUM STATE.

NATURAL SYSTEMS ARE IMMENSELY
COMPLEX, WHILE OUR UNDERSTANDING
OF THEM REMAINS LIMITED. TO SOME
EXTENT, IT'S BOUND TO STAY THAT
WAY: WE CAN RARELY DO CONTROLLED
EXPERIMENTS ON WHOLE ECOSYSTEMS
(THAT WOULD REQUIRE A SECOND,
IDENTICAL ECOSYSTEM!), AND WE
CANNOT KNOW OR OBSERVE ALL THE
RELEVANT VARIABLES.

SO WE EXPERIMENT ON SIMPLE SYSTEMS IN THE LAB, OR SIMULATE THEM ON THE
COMPUTER. THE RESULTS ARE OFTEN SUGGESTIVE, BUT WE HAVE LIMITED CONFI-
DENCE THAT THEY CAN BE APPLIED TO THE REAL WORLD.

NEVERTHELESS, WE DO
KNOW A FEW THINGS
ABOUT HOW SPECIES
INTERACT... AND IN OUR
NEXT CHAPTER WE TURN
TO ONE OF THE
SIMPLEST INTERACTIONS
OF ALL...

· CHAPTER 6 ·

ALTHOUGH YOU MAY NOT USUALLY THINK OF IT THIS WAY, THE ACT OF EATING A FELLOW CREATURE IS A WAY OF CAPTURING ITS *CHEMICAL ENERGY.*

TOFU BURGER, FRIES, LARGE GREEN COLA!

WHEN THE VEGGIE BURGER HITS YOUR STOMACH, ENZYMES GO TO WORK BREAKING DOWN ITS CHEMICAL CONSTITUENTS AND DEPLOYING THEM TO VARIOUS SYSTEMS OF YOUR BODY.

CHOFF CHOFF

THE BURGER'S FATS, FOR EXAMPLE, ARE *BURNED* IN YOUR CELLS, COMBINING WITH THE OXYGEN YOU INHALE TO RELEASE *HEAT ENERGY,* JUST LIKE A REGULAR FIRE, ONLY SLOWER.

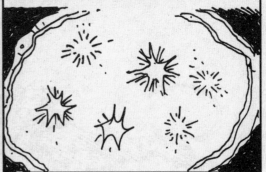

YOU USE THAT ENERGY TO WARM YOUR BODY TO 37° C, TO MOVE AROUND, TO THINK, TO BREATHE, IN SHORT, TO LIVE!

I LIVE TO EAT!

I EAT TO HEAT!

THE FOOD ALSO SERVES AS YOUR BODY'S SOURCE OF RAW MATERIALS.

HOW DO I LOOK?

LEAFY.

CHOMP
MUNCH
DISSOLVE
BITE
DIGEST
GURGLE

OF COURSE, YOU'RE NOT THE ONLY ANIMAL IN THE WORLD THAT'S EATING... THEY'RE ALL DOING IT... THE WHOLE BIOSPHERE IS BUSY MOVING ENERGY AROUND BY THE PROCESS OF EATING.

HOW THE ENERGY FLOWS THROUGH AN ECOSYSTEM DEPENDS ON *WHO EATS WHOM*. THE SEQUENCE OF DINERS AND DINNERS IS CALLED THE *FOOD CHAIN*, OR MORE ACCURATELY THE *FOOD WEB*, BECAUSE THE CHAINS ARE INTERLINKED AND MAY SHIFT AROUND.

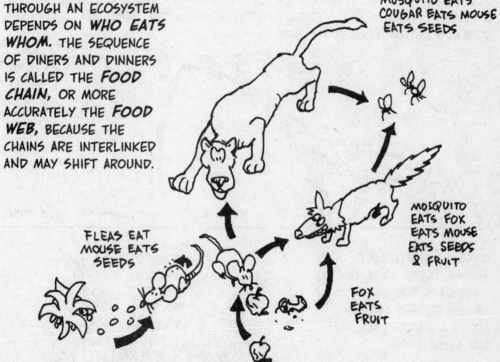

MOSQUITO EATS COUGAR EATS MOUSE EATS SEEDS

FLEAS EAT MOUSE EATS SEEDS

MOSQUITO EATS FOX EATS MOUSE EATS SEEDS & FRUIT

FOX EATS FRUIT

A GENERALIST LIKE THE FOX, FOR EXAMPLE, EATS WHATEVER IS ABUNDANT AND IN SEASON: MICE, BERRIES, GRASSHOPPERS, FALLEN APPLES...

ACCORDING TO THE

FIRST LAW OF THERMODYNAMICS,

ENERGY IS NEITHER
CREATED NOR DESTROYED.

IT JUST CHANGES
FROM ONE FORM
TO ANOTHER!

NONE OF THESE DEVOURINGS CAN CREATE ANY NEW ENERGY: THEY JUST PASS
ALONG ENERGY THAT CAME FROM SOMEWHERE ELSE. AND WHERE IS THAT? IF
YOU TRACE BACK THE LINKS OF ANY FOOD CHAIN FAR ENOUGH, YOU'LL COME TO
A **PLANT,** WHICH GOT ITS ENERGY STRAIGHT FROM THE **SUN.** ALL LIFE
DEPENDS ULTIMATELY ON THE SUN.*

THANK
YOU!

THANK
YOU!

THANK
YOU!

*WITH A FEW EXCEPTIONS. SEE NEXT PAGE.

UNTIL HUMANITY CAME
ALONG, THERE WERE ONLY
TWO WAYS FOR LIVING
BEINGS TO GET ENERGY:
SIT IN A WARM PLACE, OR
EAT SOMETHING!

WHO NEEDS
MORE?

OF THE TOTAL SUNLIGHT FALLING ON EARTH, **30%** IS REFLECTED, NEARLY **50%** IS CONVERTED TO HEAT, AND ALMOST ALL THE REST POWERS THE WATER CYCLE— EVAPORATION, RAIN, WIND, ETC. **LESS THAN 1%** IS USED BY LIVING PLANTS.

BUT THAT TINY FRACTION OF SOLAR ENERGY PROVIDES ALL OF LIFE'S FOOD NEEDS, THROUGH THE PROCESS OF

PHOTO-SYNTHESIS.

PHOTO-WHAT?

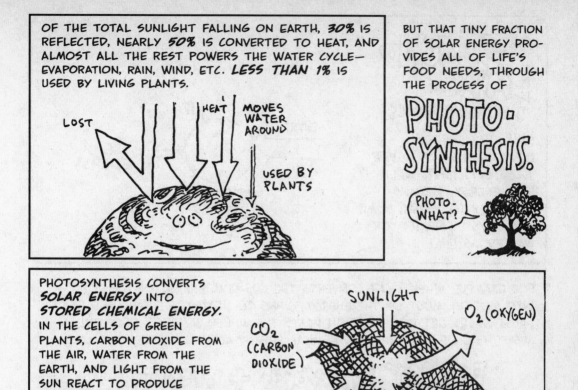

LOST

HEAT

MOVES WATER AROUND

USED BY PLANTS

PHOTOSYNTHESIS CONVERTS **SOLAR ENERGY** INTO **STORED CHEMICAL ENERGY.** IN THE CELLS OF GREEN PLANTS, CARBON DIOXIDE FROM THE AIR, WATER FROM THE EARTH, AND LIGHT FROM THE SUN REACT TO PRODUCE **SUGAR,** A COMPLEX ORGANIC COMPOUND THAT STORES CHEMICAL ENERGY FOR LATER USE. OXYGEN IS RELEASED AS A BY-PRODUCT.

SUNLIGHT

O_2 (OXYGEN)

CO_2 (CARBON DIOXIDE)

H_2O (WATER)

SUGAR

THIS STORED CHEMICAL ENERGY DRIVES ALL THE BIOGEOCHEMICAL CYCLES OF THE EARTH.

FRANKLY, I'M UNDER-UTILIZED!

EXCEPT: AT **UNDERSEA VOLCANIC VENTS,** SULFUR-LOVING BACTERIA CAN CONVERT THE **EARTH'S HEAT** INTO CHEMICAL ENERGY, A PROCESS CALLED **CHEMOSYNTHESIS.** THESE BACTERIA SUPPORT A COMMUNITY OF WORMS, CRABS, AND CLAMS IN THE TOTAL DARKNESS.

IT CAN BE DONE IN THE DARK!

VOLCANIC HEAT COMES FROM THE ENERGY OF **RADIOACTIVE DECAY** OF ELEMENTS IN THE EARTH.

THE SECOND LAW OF THERMODYNAMICS

SAYS THAT ENERGY CONVERSIONS ARE **NEVER 100% EFFICIENT:** WHENEVER ENERGY IS TRANSFORMED INTO **WORK,** SOME IS ALWAYS DISSIPATED, OR WASTED, AS HEAT.

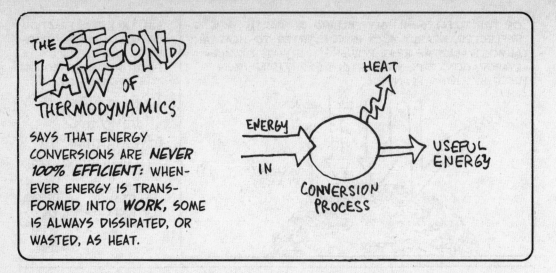

FOR EXAMPLE, WHEN A CAR CONVERTS THE CHEMICAL ENERGY OF GASOLINE INTO MOTION, MOST OF THE ENERGY TURNS TO HEAT: THE ENGINE AND EXHAUST GASES GET HOT, FRICTION HEATS THE WHEEL BEARINGS, ETC... ONLY ABOUT **15%** OF THE ORIGINAL CHEMICAL ENERGY ACTUALLY MOVES THE CAR!

ENERGY IN = WORK + HEAT
AND OTHER WASTED ENERGY

IN THE SAME WAY, **EATING** IS INEFFICIENT: ONLY A SMALL PORTION OF THE CHEMICAL ENERGY IN A MEAL CAN BE USED BY THE EATER.

ECOLOGICAL EFFICIENCY,

OR *FOOD CHAIN EFFICIENCY*, IS THE PERCENTAGE OF USABLE ENERGY CAPTURED AT EACH LEVEL OF CONSUMPTION. FOR EXAMPLE, PLANTS RANGE IN EFFICIENCY BETWEEN *1%* AND *3%* DEPENDING ON THE PLANT: ONLY 1–3% OF THE SOLAR ENERGY ABSORBED BY THE PLANT IS ACTUALLY CONVERTED TO BIOMASS.

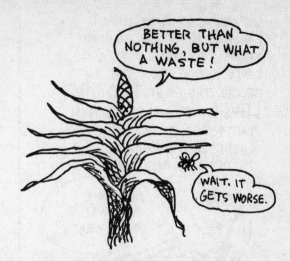

BETTER THAN NOTHING, BUT WHAT A WASTE!

WAIT. IT GETS WORSE.

A TYPICAL HERBIVORE USES SOME *10%* OF THE TOTAL PLANT ENERGY CONSUMED, WITH THE REST LOST TO HEAT OR RESPIRATION.

I NEED A LOT OF GRASS.

A CARNIVORE'S EFFICIENCY IS ALSO ABOUT *10%*, MEANING THE CARNIVORE GETS ONLY 1/10 OF 1/10 OF THE ORIGINAL PLANT ENERGY.

AND I NEED 10 TIMES AS MUCH GRASS AS YOU DO!

THE TOTAL EFFICIENCY AT ANY LEVEL OF CONSUMPTION IS THE PRODUCT OF THE EFFICIENCIES OF ALL THE CONVERTORS UP TO THAT LEVEL. IN THIS CASE (TAKING THE PLANT EFFICIENCY AS 2%), THE CARNIVORE'S TOTAL EFFICIENCY IS

$$0.02 \times 0.1 \times 0.1 = 0.0002.$$

THE CARNIVORE USES ONLY *0.02%* OF THE SOLAR ENERGY THAT WENT INTO THE GRASS THAT WENT INTO THE COW THAT WENT INTO THE CARNIVORE!

SHORTER CHAINS ARE MORE EFFICIENT!

WE CAN ALSO THINK OF THE LINKS IN THE FOOD CHAIN IN TERMS OF LEVELS, WITH EACH LEVEL EATING FROM THE ONE BELOW. THE FIRST **TROPHIC* LEVEL** CONSISTS OF GREEN PLANTS, AS WELL AS PHOTO-SYNTHESIZING AND CHEMO-SYNTHESIZING BACTERIA.

ARE YOU READY FOR MORE GREEK?

PRODUCER SPECIES ARE ALSO CALLED **AUTOTROPHS**, SINCE THEY MAKE THEIR OWN FOOD FROM CARBON DIOXIDE AND WATER.

WHO NEEDS YOU?

NONPLANT **HETEROTROPHS** CAN'T MAKE THEIR OWN FOOD, SO THEY ARE CONSUMERS OF PLANTS OR OTHER HETEROTROPHS.

CONSUMER REPORTS

*FROM GREEK *TROPHIKOS* = NOURISHMENT, AS IN ENGLISH "TROUGH."

THE SECOND TROPHIC LEVEL CONSISTS OF HERBIVORES OR PLANT-EATERS. THEY ARE ALSO CALLED **PRIMARY CONSUMERS.**

THE THIRD TROPHIC LEVEL CONSISTS OF CARNIVORES THAT EAT HERBIVORES—ALSO CALLED **SECONDARY CONSUMERS**... AND IN SOME ECOSYSTEMS THERE MAY BE LEVELS OF CARNIVORES EATING CARNIVORES: **TERTIARY** OR **QUATERNARY** CONSUMERS.

AND...

THERE ARE ALSO **OMNIVORES** THAT EAT FROM MORE THAN ONE TROPHIC LEVEL AT A TIME.

WHERE?

THESE STACKED LAYERS OF EATERS ARE **MACROCONSUMERS**, I.E., THEY'RE BIG.

BUT THOSE CONSUMERS YOU CAN'T SEE, THEY'RE THE ONES I WORRY ABOUT!

THERE ARE ALSO **MICRO-CONSUMERS**, OR DECOMPOSERS, MOSTLY BACTERIA OR FUNGI.

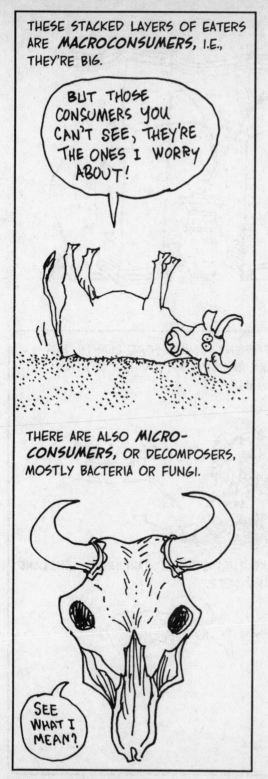

SEE WHAT I MEAN?

IN MANY TERRESTRIAL ECOSYSTEMS, AS MUCH AS 90% OF ALL PLANT MATTER FALLS STRAIGHT INTO THE DECOMPOSER FOOD WEB.

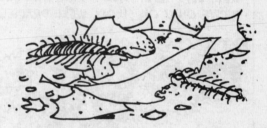

EARLY STAGES MAY INVOLVE MILLIPEDES, WEEVILS, ETC., BUT THE FINAL BREAKDOWN IS DONE BY BACTERIA AND FUNGI.

WHICH MAY RE-ENTER THE FOOD WEB IN A SOUP!

EACH SPECIES HAS TO CARVE OUT ITS PLACE IN AN ECOSYSTEM. THIS MEANS FINDING A HABITAT, TOLERABLE LEVELS OF WATER, LIGHT, AND TEMPERATURE, AND A SUPPLY OF AVAILABLE FOOD. THIS FAVORABLE COMBINATION OF FACTORS IS CALLED A

NICHE.

DIFFERENT SPECIES TEND TO OCCUPY DIFFERENT NICHES. SOME PLANTS, FOR EXAMPLE, PREFER SHADY, MOIST, POORLY DRAINED SPOTS, WHILE OTHERS THRIVE ONLY ON SUNNY, WELL-DRAINED HILLSIDES.

A NICHE CAN BE HIGHLY SPECIALIZED, LIKE THAT OF THE STARLINGS IN ENGLAND THAT EAT ONLY THE TICKS OFF SHEEP AND DEER.

WHEN RESOURCES ARE ABUNDANT, NICHES CAN BE SO BROAD THEY OVERLAP, BUT WHEN RESOURCES ARE LIMITED (THE USUAL SITUATION), OVERLAPPING NICHES MEANS **COMPETITION.**

WHEN TWO SPECIES COMPETE FOR THE SAME SCARCE RESOURCES, ONE SPECIES GENERALLY WILL *TAKE OVER THE NICHE* AND EXCLUDE THE OTHER, WHICH HAS TO FIND SOME OTHER WAY TO FEND FOR ITSELF. THIS IS THE **PRINCIPLE OF COMPETITIVE EXCLUSION.**

LET'S GO TO *PATAGONIA,* EVOLVE INTO GIANTS, AND LEARN TO EAT SEAGULLS!

IF RESOURCES PERMIT, TWO SPECIES MAY SPECIALIZE AND *PARTITION,* OR SPLIT, A NICHE INTO TWO SMALLER ONES. THE SHAG AND THE CORMORANT ARE BOTH FISH-EATING DIVING BIRDS, BUT THE CORMORANT DIVES DEEPER, LEAVING THE FISH ON TOP TO THE SHAG.

BUT IF I RUN OUT OF FISH DOWN HERE, **ONE** OF US WILL HAVE TO GO!

YEAH.

WHEN TWO SPECIES REACH SUCH AN ACCOMMODATION, THEY MAY EVOLVE **SPECIALIZED FEATURES.** FOR INSTANCE, IF TWO SPECIES OF SIMILAR BIRDS COEXIST, ONE MAY EVOLVE A LONG THIN BILL TO EAT LARGE BUGS, WHILE THE OTHER WILL HAVE A TOUGH, STUBBY BILL GOOD FOR CRACKING SEEDS. THIS IS CALLED **CHARACTER DISPLACEMENT.**

YOU'RE SO NON-THREATENING!

UNLESS YOU'RE AN INSECT!

SOMETIMES, NO ACCOMMODATION IS POSSIBLE. ONE SPECIES HAS A **COMPETITIVE ADVANTAGE:** IT'S FASTER, STRONGER, OR MORE TOLERANT OF CHANGES IN SOME LIMITING FACTORS... WITH THE RESULT THAT THE LESS WELL ADAPTED SPECIES IS DRIVEN TO COMPLETE **EXTINCTION.**

GEEZ... I'M **SO** SORRY...

NOT!

SOME LESS COMPETITIVE WAYS FOR SPECIES TO INTERACT INCLUDE:

SYMBIOSIS

("LIVING TOGETHER") IS A LONG-TERM RELATIONSHIP IN WHICH TWO SPECIES EXCHANGE ENERGY OR ADAPTIVE BENEFITS. THERE ARE THREE KINDS OF SYMBIOSIS, DEPENDING ON WHICH SPECIES RECEIVES THE MOST BENEFIT.

IN **COMMENSALISM,** ONE SPECIES GAINS, WHILE THE OTHER IS ESSENTIALLY UNAFFECTED. FOR EXAMPLE, WE EAT NUTS, BUT THE NUT TREE BARELY NOTICES.

ACTUALLY, I DO NOTICE, BUT I'M VERY BAD AT EXPRESSING MY DISAPPROVAL!

IN **MUTUALISM,** BOTH SPECIES BENEFIT. WE SAW AN EXAMPLE OF MUTUALISM IN **CORAL,** WHICH PROVIDES SHELTER TO MICRO-ORGANISMS THAT SECRETE NUTRIENTS USEFUL TO THE CORAL.

IN **PARASITISM,** THE PARASITE HARMS ITS HOST, SUCKING AWAY ITS ENERGY, USUALLY SLOWLY. EXAMPLES ABOUND: LEECHES, FLEAS, TICKS, IN-FECTIONS BY BACTERIA, PROTOZOA, FUNGI, OR WORMS, ETC..

DISGUSTING!

PREDATION

IS A RATHER SIMPLE-MINDED RELATIONSHIP: ONE SPECIES, THE PREDATOR, **HUNTS AND EATS** THE OTHER SPECIES, THE PREY.

THIS IS ONE OF THE MAIN WAYS THAT ENERGY MOVES THROUGH AN ECOSYSTEM. (DECOMPOSITION IS THE OTHER.)

ALTHOUGH PREDATION SEEMS NASTY, IT DOES BENEFIT THE PREY SPECIES. PREDATORS TEND TO TAKE THE EASY PICKINGS, THE SLOW AND INFIRM, LEAVING THE MOST FIT AND VIGOROUS TO SURVIVE.

LIONS ATE ALL THE LOW JUMPERS!

PREDATOR AND PREY POPULATIONS TEND TO BE IN **DYNAMIC BALANCE.** OVERPREDATION REDUCES PREY POPULATIONS, SO PREDATORS DIE OF STARVATION, ALLOWING PREY POPULATIONS TO REBOUND. THEN PREDATORS FLOURISH, OVEREAT AGAIN, AND DIE OFF, ETC. A FAMOUS GRAPH BY LOTKA-VOLTERRA SHOWS HOW THESE TWO POPULATIONS RISE AND FALL WITH EACH OTHER, SEPARATED BY A TIME LAG.

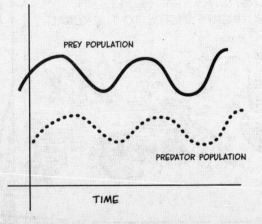

PREY POPULATION

PREDATOR POPULATION

TIME

THE PREDATOR SPECIES PROWL THE TOP OF THE FOOD CHAIN. THE FLOW OF FOOD ENERGY MOVES UPWARD FROM PLANT TO PREDATOR, AND THERE IT STOPS. (ON LAND, PREDATORS RARELY PREY ON OTHER PREDATORS; UNDERWATER, THERE MAY BE MORE TROPHIC LEVELS, BUT THEY HAVE TO END SOMEWHERE.)

FOOD ENERGY ARROWS GO INTO US AND DON'T COME OUT!

OF COURSE, A LITTLE ENERGY FLOWS FROM PREDATORS TO THEIR PARASITES (FLEAS, TICKS, SHARK-SUCKING LAMPREYS, INTESTINAL BACTERIA, ETC.) BUT IT ISN'T MUCH.

SEEMS LIKE A LOT TO US!

ONLY AT DEATH DOES THE PREDATOR'S CHEMICAL ENERGY RETURN TO THE ECOSYSTEM VIA THE DECOMPOSERS AND SCAVENGERS.

EVEN THE PREDATOR DECOMPOSES IN THE END...

AMONG THE LIMITING FACTORS FOR PREDATORS ARE THESE:

THE SUPPLY OF PREY.

MOVE!

I CAN'T

THE PREDATOR'S HUNTING SKILL.

OOF! WAIT! HOLD ON! RRF!

HAVE YOU CONSIDERED A SWITCH TO CARRION?

THE PREDATOR'S VERSATILITY AT SHIFTING ITS DIET WHEN NECESSARY.

I'D RATHER STARVE THAN EAT NAKED MOLE RAT!

AND THE ULTIMATE LIMIT, THE TOTAL ENERGY AVAILABLE AT THE PREDATOR'S TROPHIC LEVEL.

NOW TRY TO IMAGINE A *SUPERPREDATOR* THAT COULD SOMEHOW *OVERCOME* THOSE LIMITING FACTORS... A SPECIES THAT CONTINUALLY *IMPROVES ITS HUNTING TECHNIQUE* AND IS WILLING TO EAT NEARLY *ANYTHING*... AND WHAT IF THIS SPECIES ALSO DISCOVERED *SECRET SOURCES OF ENERGY* THAT HAD ALWAYS BEEN HIDDEN FROM ALL OTHER ANIMALS? TRY TO IMAGINE IT...

BY GAIA!! WHERE IS IT??

· CHAPTER 7 ·

FROM HUNTING TO PLANTING

LIKE ALL CREATURES, WE ARE PART OF THE WEB OF LIFE. WE EAT FOOD FROM EVERY **TROPHIC LEVEL**; WE HAVE FEW **PREDATORS**; AND WE HAVE MANY **SYMBIOTIC** RELATIONSHIPS: WE PRACTICE **MUTUALISM** WITH THE BACTERIUM *E. COLI*, WHICH HELPS US DIGEST FOOD IN EXCHANGE FOR A SAFE, WARM HOME IN OUR INTESTINES; WE HAVE **PARASITIC** DISEASES; AND WE HAVE **COMMENSAL** RELATIONSHIPS WITH PIGEONS AND COCKROACHES (THEY THRIVE; WE'RE UNAFFECTED).

OF COURSE, OUR WAY OF LIFE HAS CHANGED OVER THE YEARS. IN THE BEGINNING, SOME 2 MILLION YEARS AGO, OUR ANCESTORS OCCUPIED THE NICHE OF **HUNTER-GATHERERS,** WHICH REMAINED HUMANITY'S NICHE FOR 99% OF OUR HISTORY.

STUDIES OF MODERN SOCIETIES SUGGEST THAT HUNTING AND GATHERING IS SURPRISINGLY EASY. FOOD IS ABUNDANT IF YOU KNOW WHAT TO LOOK FOR, SO HUNTER-GATHERERS DON'T HAVE TO WORK VERY HARD. THEY HAVE PLENTY OF TIME FOR PLAYING, PERFORMING RITUALS, OR DISCUSSING FAMILY ISSUES.

HUNTER-GATHERER POPULATIONS ARE SMALL, RARELY EXCEEDING 50... THEY HAVE TO BE MOBILE, SO POSSESSIONS ARE A BURDEN... AND POPULATION CONTROL IS PRACTICED IN THE FORMS OF INFANTICIDE, ABANDONMENT OF THE AGED AND INFIRM, AND EVEN MURDER.

IF A GROUP GROWS TOO LARGE FOR ITS TERRITORY TO SUSTAIN IT, A FACTION SPLITS OFF AND MOVES AWAY.

WHAT CHANGED THIS STABLE IF NOT ENTIRELY IDYLLIC PICTURE? SURELY IT WAS THE MASTERY OF

FIRE.

WELL, IT'S NICE, BUT WHAT'S THE BIG DEAL?

FIRE WAS THE FIRST OF THE *HIDDEN ENERGY SOURCES* LIBERATED BY HUMANS. WHAT ANIMAL (ASIDE FROM TERMITES) EVER THOUGHT OF GETTING USEFUL ENERGY FROM *WOOD?* WITH THIS EXTRA, PORTABLE, NONFOOD ENERGY, PEOPLE COULD SUDDENLY LIVE ANYWHERE. CLIMATE WAS NO BARRIER...

THERE GOES ONE LIMITING FACTOR!

(ESPECIALLY AFTER THEY INVENTED *CLOTHES*, THE FIRST ENERGY-CONSERVATION TECHNOLOGY*).

I CALL IT "INSULATION!"

*OR MAYBE THE SECOND. HOUSES ALSO RETAIN HEAT.

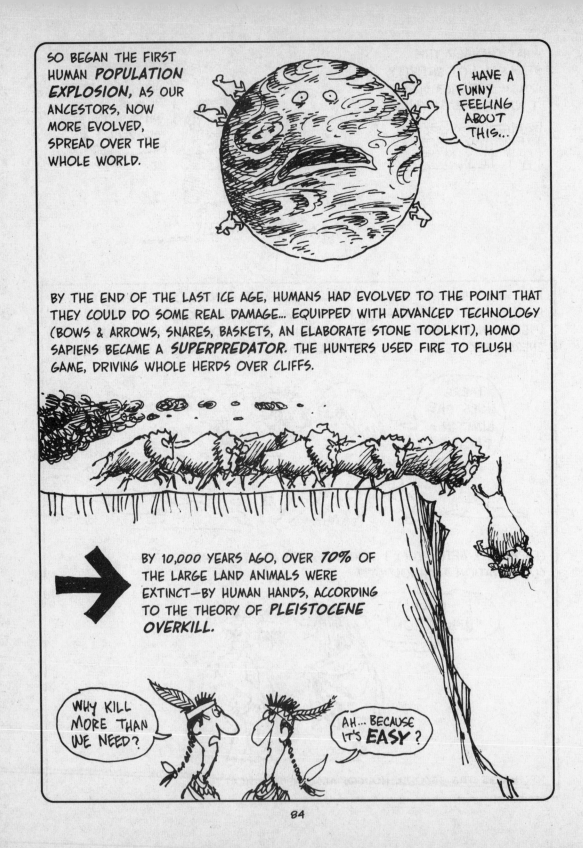

SO BEGAN THE FIRST HUMAN **POPULATION EXPLOSION**, AS OUR ANCESTORS, NOW MORE EVOLVED, SPREAD OVER THE WHOLE WORLD.

I HAVE A FUNNY FEELING ABOUT THIS...

BY THE END OF THE LAST ICE AGE, HUMANS HAD EVOLVED TO THE POINT THAT THEY COULD DO SOME REAL DAMAGE... EQUIPPED WITH ADVANCED TECHNOLOGY (BOWS & ARROWS, SNARES, BASKETS, AN ELABORATE STONE TOOLKIT), HOMO SAPIENS BECAME A **SUPERPREDATOR**. THE HUNTERS USED FIRE TO FLUSH GAME, DRIVING WHOLE HERDS OVER CLIFFS.

BY 10,000 YEARS AGO, OVER **70%** OF THE LARGE LAND ANIMALS WERE EXTINCT—BY HUMAN HANDS, ACCORDING TO THE THEORY OF **PLEISTOCENE OVERKILL.**

WHY KILL MORE THAN WE NEED?

AH... BECAUSE IT'S **EASY** ?

NOT EVERYONE ACCEPTS THE THEORY OF PLEISTOCENE OVERKILL. THESE EXTINCTIONS ALSO COINCIDED WITH THE END OF THE LAST ICE AGE, AND MANY SCIENTISTS CONTEND THAT THE ANIMALS DIED AS A RESULT OF CLIMATIC CHANGE. BUT IT IS HARD TO IMAGINE THAT PEOPLE PLAYED NO ROLE IN TIPPING THE BALANCE AGAINST ALL THOSE MAMMOTHS, GROUND SLOTHS, GIANT ELK, ETC.

THIS MUST HAVE BEEN A SERIOUS CRISIS FOR HUMANITY: WITH A POPULATION IN THE NEIGHBORHOOD OF 4 MILLION, OUR SPECIES WAS SUDDENLY FACED WITH A SHORTAGE OF BIG GAME...

BUT OUR ANCESTORS WERE INTELLIGENT, SOCIAL OMNIVORES WITH SOME TECH-NOLOGY, AND THEY WERE ABLE TO TURN A CRISIS INTO AN OPPORTUNITY.

THEY INVENTED AGRICULTURE, USHERING IN THE MOST FUNDAMENTAL UPHEAVAL IN HUMAN HISTORY, AND POSSIBLY IN ALL WORLD HISTORY.

WHAT IS AGRICULTURE? IN TERMS OF THE FOOD WEB, FARMING MEANS A RADICAL **SIMPLIFICATION** AND **REORGANIZATION** OF THE FLOW OF ENERGY. THE FARMER UPROOTS (OR CHASES AWAY, OR BURNS OFF) UNWANTED SPECIES AND SOWS ONLY THOSE CROPS THAT ARE FIT FOR HUMAN CONSUMPTION.

IN THIS WAY, ALL (OR NEARLY ALL) THE PLANT ENERGY ON A GIVEN PLOT OF GROUND NOW FLOWS INTO HUMAN MOUTHS AT THE EXPENSE OF THE OTHER SPECIES THAT USED TO FIND A NICHE THERE.

WITH THE INVENTION OF FARMING, HUMAN SOCIETY EXPERIENCED A SUDDEN INFLUX OF ENERGY. THIS HAD TWO MAJOR EFFECTS: **POPULATION GROWTH** AND **SOCIAL ORGANIZATION.**

WELL, WE HAVE TO DO *SOMETHING* WITH ALL THIS ENERGY!

POPULATION GREW BECAUSE AGRICULTURE INCREASES THE CARRYING CAPACITY OF THE LAND: THERE WAS MORE FOOD FOR HUMANS, HENCE MORE HUMANS!

ARE WE SURE THIS IS WHAT WE WANT?

UNLIKE THE HUNTER-GATHERER, THE FARMER WANTS A LARGE FAMILY. MORE CHILDREN MEAN MORE FIELD HANDS, HENCE MORE FOOD, NOT LESS! (THIS MENTALITY IS ONE OF THE OBSTACLES TO POPULATION CONTROL IN MAINLY RURAL COUNTRIES TODAY.)

WAIT... IF MORE PEOPLE MAKE MORE FOOD... AND MORE FOOD MAKES MORE PEOPLE.... THEN — ??

IN GENERAL, WHENEVER ANY COMPLEX SYSTEM HAS A SUPPLY OF ENERGY COMING IN, THE SYSTEM MAY TEND TO **ORGANIZE ITSELF.** (THIS BEAUTIFUL IDEA IS DUE TO THE NOBEL PRIZE-WINNING CHEMIST ILYA PRIGOGINE.)

YUP! TIME TO GET ORGANIZED!

THE INFLUX OF FOOD ENERGY FROM AGRICULTURE HAD A SIMILAR EFFECT WHEREVER IT HAPPENED: DIVISION OF LABOR... SOCIAL HIERARCHIES... GOVERNMENT COMMITTEES... TEMPLES... MARKETS...

O.K.... YOU DIG THE HOLE... YOU CARRY THE POST... YOU BRING THE LUNCH... YOU...

AND YOU? I'LL TELL *YOU* WHAT TO DO...

AND ULTIMATELY, IN SOME CASES,

CIVILIZATION.

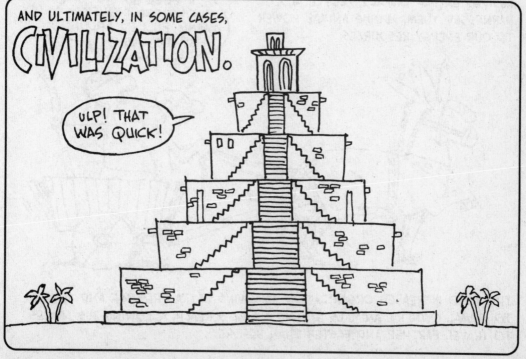

ULP! THAT WAS QUICK!

IN MOST PLACES WHERE AGRICULTURE DEVELOPED, PEOPLE ALSO DOMESTICATED ANIMALS. THIS HAD A SIMILAR EFFECT: A HERDER'S FLOCK USES MOST OF THE AVAILABLE PLANT ENERGY, LEAVING LESS FOR WILD ANIMALS, INCLUDING PREDATORS.

DOMESTICATED PLANTS AND ANIMALS BOTH LIVE IN SYMBIOTIC RELATIONSHIPS WITH PEOPLE: WE PROTECT THEM, THEY FEED US.

BESIDES EATING ANIMALS, PEOPLE ALSO HARNESSED THEM, ADDING ANIMAL POWER TO OUR ENERGY RESOURCES.

THIS ALSO INCREASED ORGANIZATION, BY GIVING EXTRA PRESTIGE AND POWER TO ANIMAL OWNERS. ANIMALS EXTEND A KING'S REACH, ALLOWING HIS ARMIES TO TRAVEL FARTHER AND FASTER THAN BEFORE.

THE FIRST FARMERS LIVED IN THE HILLS OF WESTERN ASIA AROUND 8000 B.C., WHERE THEY DOMESTICATED SHEEP AND GOATS AND BRED BETTER STRAINS OF WHEAT AND BARLEY.

SOMEWHAT LATER, THE CHINESE MASTERED MILLET, PIGS, POULTRY, AND LATER STILL, RICE.

LIFE IS SWEET... AND SOUR!

MEXICO AND ITS NEIGHBORS, GROWING *CORN* (MAIZE), WERE HANDICAPPED BY THE PLANT'S GENETIC PECULIARITIES FROM BREEDING HIGH-YIELD VARIETIES UNTIL 2000 B.C. MEANWHILE, THEY BRED PEPPERS, SQUASH, TOMATOES, AND CHOCOLATE, WHILE THE PERUVIANS PLANTED POTATOES AND LOADED LLAMAS...

WHILE AFRICANS PLANTED YAMS AND SORGHUM AND RAISED CATTLE.

BY *2000* B.C., VIRTUALLY EVERYTHING WE NOW FARM HAD BEEN DOMESTICATED. ONLY ABOUT *50 ANIMAL SPECIES* HAVE EVER BEEN TAMED (THE OTHERS CAN'T STAND IT), AND EVEN TODAY THE BULK OF THE WORLD RELIES ON A VERY LIMITED NUMBER OF STAPLE VEGETABLE FOODS.

WHEREVER PEOPLE FARMED, THEY ENCOUNTERED SOME UNFORESEEN PROBLEMS.

SUDDENLY, NOTHING'S COMING UP!

THE MOST BASIC PROBLEM IS THAT FARMING IS **EXTRACTIVE:** NUTRIENTS COME OUT OF THE SOIL BUT DON'T CYCLE BACK, AS IN WILD ECOSYSTEMS.

THROUGHOUT HISTORY, FARMERS HAVE DEALT WITH SOIL DEPLETION. THE ORIGINAL WAY WAS SIMPLY TO MOVE TO GREENER PASTURES.

LEAVE IT TO THE SUCCESSIONAL SPECIES!

WHEN THAT BECAME IMPRACTICAL, THEY FOUND WAYS TO PUT NUTRIENTS INTO THE LAND: SPREADING MANURE RETURNS PHOSPHORUS, AND PLANTING LEGUMES (PEAS, BEANS, LENTILS) RETURNS NITROGEN TO DEPLETED SOILS.

A FINE, FILTHY JOB THIS TURNED OUT TO BE!

ANOTHER TACTIC IS **INTERPLANTING:** GROWING DIFFERENT SPECIES OR VARIETIES IN THE SAME PLOT. IN CENTRAL AMERICA, THE COMBINATION WAS SQUASH, CORN, AND BEANS.

CORN SUPPORTS BEAN STALK.

SQUASH ROOTS ANCHOR SOIL.

BEANS FIX NITROGEN

WHATEVER THE TECHNIQUE, SUCCESSFUL FARMING DEPENDS IN THE FIRST PLACE ON GOOD SOIL MANAGEMENT.

ANOTHER AGRICULTURAL RISK IS EROSION, OR LOSS OF TOPSOIL.

HM... THE TOPSOIL IS HEADING FOR THE BOTTOM LANDS...

IN WILD ECOSYSTEMS, MANY PLANTS, AND ESPECIALLY TREES, COMBINE TO HOLD WATER, PRODUCE AND ENRICH SOIL, AND ANCHOR IT WITH THEIR ROOTS.

BUT SO HARD TO GET A MEAL!

CLEARED LAND DRIES UP AND IS EASILY BLOWN AWAY BY WIND OR WASHED OFF BY FLOODS.

!

THIS IS WHY CAREFUL HILLSIDE FARMERS BUILD *TERRACED FIELDS* WATERED BY CONTROLLABLE DRAINAGE SYSTEMS.

NOTE FOREST RETAINED ON RIDGE AS A SOURCE OF TRICKLE-DOWN NUTRIENTS.

DESPITE EVERYONE'S BEST EFFORTS, FARMING COMMUNITIES CAN STILL DEGRADE THEIR ENVIRONMENTS IN THE LONG RUN. WE OFFER THREE EXAMPLES, TWO ANCIENT AND ONE MODERN.

SUMER & SALT

THE FIRST GREAT CIVILIZATION, THE **SUMERIAN,** AROSE IN THE RICH SOIL BETWEEN THE TIGRIS AND EUPHRATES RIVERS, A BROAD, ALLUVIAL PLAIN.

JUST ONE QUESTION: HOW DO WE WATER IT?

THE SUMERIANS DUG COUNTLESS CANALS AND WATERED THEIR FIELDS BY CONTROLLED FLOODING, OR IRRIGATION.

AS WATER FLOWS, IT PICKS UP SALTS FROM THE SOIL, AND WHEN IT SITS STAGNANT, MUCH WATER EVAPORATES, LEAVING THE SALTS BEHIND.

LOOKIN' GOOD!

SALT

SALTY SOIL STUNTS PLANTS, REDUCES YIELDS, AND EVENTUALLY MAKES A FIELD INFERTILE.

UNLESS FIELDS ARE WELL FLUSHED FROM TIME TO TIME, SALT BUILDS UP IN THE SOIL.

FLUSH

IT'S THE RINSE CYCLE KICKING IN...

AT FIRST, SUMERIAN FARMERS SIMPLY ABANDONED SALTY PLOTS AND PLOWED ELSEWHERE...

BUT AS POPULATION SWELLED, IRRIGATION WORKS COVERED THE LAND. DIVERTING WATER TO FLUSH OUT SALTY FIELDS WOULD HAVE DISPLACED TOO MANY PEOPLE.

MEANWHILE, NEARBY HILLS WERE STRIPPED OF TREES FOR FUEL AND LUMBER, ALLOWING SOIL TO WASH DOWN TO THE VALLEY.

IRRIGATION DITCHES BECAME CLOGGED WITH SILT, AND WATER NO LONGER FLOWED.

AFTER MORE THAN *2000 SUCCESSFUL YEARS* OF FERTILITY, SUMER'S SOIL BECAME ALMOST USELESS. CROP YIELDS PLUNGED, AND BY 1700 B.C., POPULATION HAD FALLEN BY A FACTOR OF 10, TO AROUND 150,000.

THE LAND NEVER RECOVERED. IT IS NOW THE SCRUBBY DESERT OF SOUTHERN IRAQ.

MEXICAN PHOSPHORUS

IN CENTRAL AMERICA, CORN FED THE MAYAN PEOPLE. AT ITS HEIGHT IN 800 A.D., THE CITY OF **TEOTIHUACAN** COVERED 10 SQUARE MILES AND HAD AT LEAST 100,000 INHABITANTS.

BUT THE SEEDS OF ITS RUIN WERE ALREADY SOWN. THE DEFORESTATION OF MEXICO WAS COMPLETED AROUND THE YEAR 250.

THE CORN IS AS *HIGH*... AS... AS... A...

DESPITE ALL THE MAYAS' EFFORTS AT WATER CONTROL, TERRACING, AND RAISED FIELDS (FOR BETTER DRAINAGE), EROSION WENT STEADILY ON.

AS A CHICKEN'S LARYNX?

PUK

CORE SAMPLES FROM LAKE BEDS TELL THE STORY:

SOIL NUTRIENTS, ESPECIALLY PHOSPHORUS, DRAINED INTO THE LAKES. THIS WAS BAD FOR BOTH SOIL AND LAKE.

(AND THE MAYAN PEOPLE HAD NO DOMESTICATED ANIMALS WHOSE MANURE COULD REPLENISH THE POOR SOIL.)

WITHIN 100 YEARS OR SO OF ITS GREATEST GLORY, MAYAN CIVILIZATION WAS GONE, ITS SHRUNKEN POPULATION REDUCED TO HACKING LITTLE FARM PLOTS OUT OF THE JUNGLE.

96

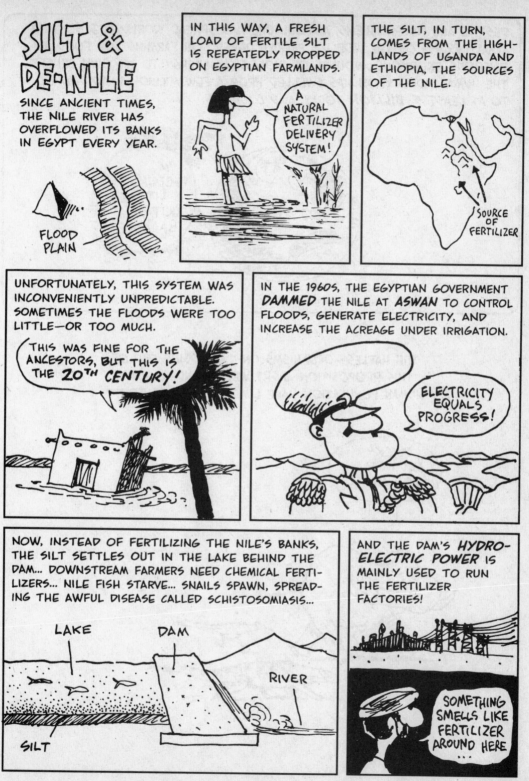

SILT & DE-NILE

SINCE ANCIENT TIMES, THE NILE RIVER HAS OVERFLOWED ITS BANKS IN EGYPT EVERY YEAR.

FLOOD PLAIN

IN THIS WAY, A FRESH LOAD OF FERTILE SILT IS REPEATEDLY DROPPED ON EGYPTIAN FARMLANDS.

A NATURAL FERTILIZER DELIVERY SYSTEM!

THE SILT, IN TURN, COMES FROM THE HIGHLANDS OF UGANDA AND ETHIOPIA, THE SOURCES OF THE NILE.

SOURCE OF FERTILIZER

UNFORTUNATELY, THIS SYSTEM WAS INCONVENIENTLY UNPREDICTABLE. SOMETIMES THE FLOODS WERE TOO LITTLE—OR TOO MUCH.

THIS WAS FINE FOR THE ANCESTORS, BUT THIS IS THE 20TH CENTURY!

IN THE 1960s, THE EGYPTIAN GOVERNMENT DAMMED THE NILE AT ASWAN TO CONTROL FLOODS, GENERATE ELECTRICITY, AND INCREASE THE ACREAGE UNDER IRRIGATION.

ELECTRICITY EQUALS PROGRESS!

NOW, INSTEAD OF FERTILIZING THE NILE'S BANKS, THE SILT SETTLES OUT IN THE LAKE BEHIND THE DAM... DOWNSTREAM FARMERS NEED CHEMICAL FERTILIZERS... NILE FISH STARVE... SNAILS SPAWN, SPREADING THE AWFUL DISEASE CALLED SCHISTOSOMIASIS...

LAKE   DAM

RIVER

SILT

AND THE DAM'S HYDRO-ELECTRIC POWER IS MAINLY USED TO RUN THE FERTILIZER FACTORIES!

SOMETHING SMELLS LIKE FERTILIZER AROUND HERE ...

DESPITE ALL ITS PROBLEMS, AGRICULTURE HAS SPREAD WORLDWIDE, SO THAT TODAY *99.9%* OF THE WORLD'S PEOPLE DEPEND ON FARMING FOR FOOD. BY DIVERTING ENORMOUS AMOUNTS OF BIOLOGICAL ENERGY TO ITS OWN BENEFIT, THE HUMAN POPULATION HAS SWELLED FROM A FEW MILLION IN 10,000 B.C. TO AT LEAST *5 BILLION* TODAY, AND COUNTING.

THE HAPLESS ORGANISMS ON THE LOSING END OF THIS PROPOSITION MIGHT WELL WONDER: DON'T HUMAN POPULATION HAVE LIMITING FACTORS TOO??

• CHAPTER 8 •

WHAT LIMITING FACTORS?

HOW DO POPULATIONS GROW?
WHY ARE THERE SO MANY PEOPLE?
AND WHY AREN'T THERE EVEN MORE?

BEFORE EXPLORING THE EXPANSION OF THE HUMAN SPECIES, CONSIDER THIS PARADOX: AFTER THE INVENTION OF AGRICULTURE, MOST PEOPLE WERE **WORSE OFF** THAN THEIR ANCESTORS. FROM SKELETONS WE KNOW THAT WHEN AGRICULTURE ARRIVED, AVERAGE HEIGHT DECLINED ABOUT 4 INCHES, FROM 5'9" TO 5'5" FOR MEN AND FROM 5'5" TO 5'1" FOR WOMEN. COMPARED TO HUNTER-GATHERERS, FARMERS WORK HARDER, EAT A LESS NUTRITIOUS DIET, AND GET SICK MORE OFTEN.

A SYSTEM, AGRICULTURE, THAT MAKES PEOPLE SMALL AND SICKLY, HAS SUPPLANTED A SYSTEM, HUNTING AND GATHERING, THAT MAKES PEOPLE HEALTHY AND ROBUST. HOW CAN THIS BE?

THE ANSWER LIES IN AGRICULTURE'S **SYSTEM DYNAMICS:** ONCE IT BEGINS, FARMING TENDS TO EXPAND!

OH, GREAT!

IN THE FIRST PLACE, AGRICULTURE GENERALLY PRODUCES A **SURPLUS:** AT HARVEST TIME, THERE ARE HEAPS OF GRAIN OR GOURDS OR POTATOES—TOO MUCH TO EAT ON THE SPOT, SO IT'S PUT INTO STORAGE FOR THE WINTER. MOST OF THE COMMUNITY'S FOOD IS IN ONE PLACE AT ONE TIME.

AT LEAST, I HOPE IT'S A SURPLUS!

NOW THE QUESTION ARISES, WHO OWNS THAT SURPLUS?

NOT ME.

NOT ME.

I DO!

UNLIKE HUNTING AND GATHERING, AGRICULTURE ENABLES THE STRONGEST PEOPLE IN THE COMMUNITY TO LAY HANDS ON EVERYONE'S ENERGY RESOURCES AND CONTROL THEM.

WHO SAID AGRICULTURE WASN'T GOOD FOR PEOPLE?

IN OTHER WORDS, WITH AGRICULTURE COMES A **LANDLORD CLASS:** KINGS, PRIESTS, AND NOBLES, WHO TAKE A CUT OF THE HARVEST FOR THEMSELVES, AND NOT A SMALL ONE EITHER—50% IS TYPICAL!

FAIR ENOUGH?

FAIR TO POOR...

THAT CONCENTRATED SURPLUS IS THE SOURCE OF **SOCIAL ORGANIZATION.** THE LANDLORDS USE THEIR WEALTH TO HIRE ARMIES, BUILD ROADS, AND HIRE THINKERS TO DREAM UP IDEAS TO KEEP THE PEASANTS IN LINE.

LISTEN TO THIS ONE— I CALL IT "THE DIVINE RIGHT OF KINGS..."

GREAT STUFF! GREAT STUFF!

THE WHOLE POINT OF THIS EXERCISE CALLED CIVILIZATION IS TO KEEP THE SURPLUSES OF ENERGY—STARTING WITH FOOD ENERGY—FLOWING INTO THE HANDS OF THE STRONG. IN OTHER WORDS, IT WAS ORGANIZED FROM THE OUTSET TO BE **SELF-PERPETUATING.**

I WOULDN'T HAVE IT ANY OTHER WAY!

THEN, TOO, WE'VE SEEN HOW THE LOGIC OF FARMING COMPELS THE POOR TO HAVE MORE CHILDREN RATHER THAN FEWER: MORE HANDS MEANS MORE WORK CAN BE DONE.

AND IF YOU SHOULD WANT TO STOP WORKING AND GO BACK TO HUNTING, FORGET IT! HUNTING WAS STRICTLY REGULATED, EVEN FORBIDDEN TO COMMONERS. THIS IS A SPORT FOR THE NOBLES!

THE CONSEQUENCE IS **RAPID POPULATION GROWTH.** THE NEW GENERATIONS CLAIM MORE LAND FOR FARMING... TRADITIONAL PEOPLES, WITH THEIR POPULATION CONTROLS, ARE OUTNUMBERED AND OVER-POWERED... NO WONDER AGRICULTURE TOOK OVER THE WORLD!

IN GENERAL, ANY POPULATION OF ORGANISMS, IF UNCHECKED, TENDS TO GROW *EXPONENTIALLY*. WHAT DOES THAT MEAN?

LIKE, RILLY, RILLY *FAST*?

IT MEANS THAT THE POPULATION'S *RATE OF INCREASE* IS *CONSTANT*. IN SYMBOLIC TERMS, WE CALL P_0 THE POPULATION TODAY AND r THE RATE OF INCREASE. THEN A YEAR FROM NOW THE POPULATION WILL BE $P_0 + rP_0$. A YEAR LATER, WITH r REMAINING THE SAME, THE POPULATION WILL BE $(P_0 + rP_0) + r(P_0 + rP_0)$. AFTER A TINY BIT OF ALGEBRA, THIS BECOMES $P_0(1+r)^2$, AND WE CAN GO ON TO WRITE:

POPULATION TODAY........ P_0

" AFTER **1** YEAR.... $P_0(1+r)$

" " **2** YEARS.... $P_0(1+r)^2$

" " **3** YEARS.... $P_0(1+r)^3$

" " n " .... $P_0(1+r)^n$

"EXPONENTIAL" BECAUSE THE *EXPONENT* GOES UP WITH TIME!

DRAW A GRAPH OF THIS *EXPONENTIAL FUNCTION*, AND YOU SEE THAT ITS LEFT END LOOKS QUITE DIFFERENT FROM ITS RIGHT END. AT FIRST IT GROWS ONLY SLOWLY, ALMOST IMPERCEPTIBLY, BUT AT SOME POINT IT TAKES OFF. (HERE ARE TWO EXAMPLES, ONE WITH LOW r AND ONE WITH HIGH r.)

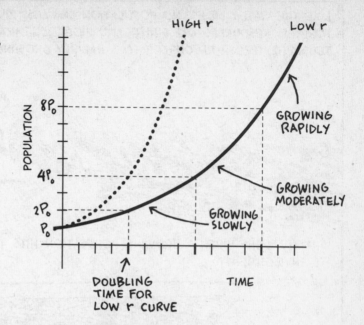

IF YOU AREN'T COMFORTABLE WITH THE ALGEBRA, ALL YOU NEED TO REMEMBER IS THIS: EXPONENTIAL GROWTH HAS A FIXED

DOUBLING TIME.

THERE IS A CERTAIN NUMBER OF DAYS, OR MONTHS, OR YEARS (IT DEPENDS ON THE SPECIES) IN WHICH THE POPULATION DOUBLES.

AN EXAMPLE IS A POND WITH A LILY PAD WHOSE DOUBLING TIME IS ONE WEEK.

YOU MAY NOT EVEN NOTICE ITS GROWTH UNTIL THE POND IS HALF COVERED.

THE NEXT WEEK, THE POND HAS DISAPPEARED UNDER LILIES!

CHOKE!

LIKE THE WATER LILIES, NO POPULATION CAN CONTINUE EXPONENTIAL GROWTH FOREVER: RESOURCES ARE FINITE, AND BIOGEOCHEMICAL CYCLES ARE TOO SLOW TO SUPPLY FRESH RESOURCES TO A RAPIDLY GROWING POPULATION.

WHAT HAPPENS WHEN A GROWING POPULATION HITS THE CARRYING CAPACITY OF ITS ENVIRONMENT?

WITH ANY LUCK, IT WILL SLOW AS IT APPROACHES THE CARRYING CAPACITY AND LEVEL OFF, REACHING A HAPPY ACCOMMODATION WITH THE ENVIRONMENT. THIS GIVES AN **S**-CURVE.

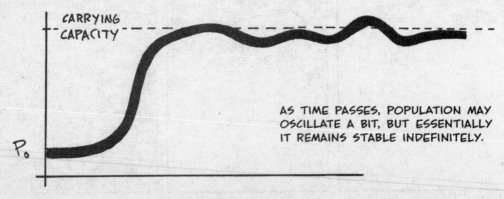

AS TIME PASSES, POPULATION MAY OSCILLATE A BIT, BUT ESSENTIALLY IT REMAINS STABLE INDEFINITELY.

A LESS ROSY SCENARIO IS DESCRIBED BY THE

J-CURVE,

WHICH LOOKS LIKE THIS: A PERIOD OF RAPID POPULATION INCREASE FOLLOWED BY COLLAPSE.

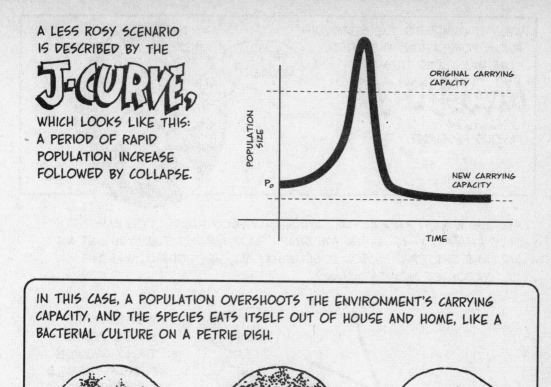

IN THIS CASE, A POPULATION OVERSHOOTS THE ENVIRONMENT'S CARRYING CAPACITY, AND THE SPECIES EATS ITSELF OUT OF HOUSE AND HOME, LIKE A BACTERIAL CULTURE ON A PETRIE DISH.

AFTER DEPLETING THE AVAILABLE RESOURCES (SUGAR IN THIS CASE), THE BACTERIA DIE OFF RAPIDLY, LEAVING ONLY A PATHETIC REMNANT OF THE ORIGINAL POPULATION.

WHEN IT COMES TO THE STUDY OF HUMAN POPULATION, THE FIRST NAME HAS TO BE THOMAS **MALTHUS** (1766-1834), FAMOUS PESSIMIST.

WE'RE DOOMED!

MALTHUS IS BEST KNOWN FOR SAYING THAT FOOD PRODUCTION CAN RISE ONLY *LINEARLY*—I.E., WHEN YOU GRAPH FOOD AGAINST TIME, YOU GET A STRAIGHT LINE. (THE REASON IT GROWS AT ALL, HE THOUGHT, WAS THE CULTIVATION OF NEW FARMLAND.)

YUP, DOOMED!

INEVITABLY, HE SAID, EXPONENTIALLY GROWING POPULATION WILL OUTSTRIP THE AVAILABLE FOOD AT SOME POINT...

BUT MALTHUS WAS NOT COMPLETELY GLOOMY! HE OPTIMISTICALLY MAINTAINED THAT HUMAN NUMBERS WERE HELD IN BALANCE WITH THE FOOD SUPPLY BY THREE "PREVENTATIVE CHECKS."

PESTILENCE, FAMINE, AND **WAR.**

SO THERE *IS* HOPE!

IT TURNS OUT THAT MALTHUS WAS WRONG ABOUT THE FOOD SUPPLY. FOR REASONS WE'LL DISCUSS IN A LATER CHAPTER, FARMERS HAVE INCREASED THEIR TOTAL OUTPUT AT A MUCH FASTER RATE THAN A STRAIGHT LINE.

OH, WELL...

HE WAS RIGHT, HOWEVER, TO IDENTIFY PESTILENCE, FAMINE, AND WAR AS FACTORS THAT LIMIT HUMAN POPULATIONS... SO LET'S LOOK AT EACH IN TURN.

WAR

HAS BEEN WITH US SINCE THE DAWN OF HISTORY. IT REDUCES POPULATION IN AN OBVIOUS AND DIRECT WAY: BY KILLING PEOPLE.

SO WHAT'S YOUR POINT?

BUT GROUPS WAGE WARS TO WREST RESOURCES FROM EACH OTHER. THE WINNING SIDE, DESPITE ITS BATTLE LOSSES, MAY GAIN ENOUGH TO THRIVE AFTERWARD, INCREASING POPULATION.

BESIDES, BIRTHRATES USUALLY GO UP AFTER WARS, WHEN SOLDIERS COME HOME TO THEIR WIVES, AND POPULATIONS REBOUND RAPIDLY.

LET'S MAKE A BABY BOOM!!

WAR DOES MOST OF ITS DAMAGE INDIRECTLY, BY SPREADING DISEASE AND FAMINE.

AN ARMY MARCHES ON ITS STOMACH, AND IN THE DAYS BEFORE CANNED AND FREEZE-DRIED RATIONS, THIS MEANT STRIPPING THE COUNTRYSIDE.

FOOD SHORTAGES LEAVE PEOPLE SUSCEPTIBLE TO DISEASE, AND ARMIES OFTEN BRING HOME FOREIGN INFECTIONS.

OF COURSE, ONE MAY PERFECTLY WELL STARVE WITHOUT A WAR. **FAMINE** HAS BEEN A PART OF AGRICULTURE FROM EARLY TIMES.

EVEN IN PEACETIME, FARMERS HAVE BEEN BADLY NOURISHED COMPARED TO THEIR HUNTER-GATHERER COUSINS. WHEN THE HARVEST IS BAD, THE EFFECT IS MALNUTRITION OR STARVATION.

(BY CONTRAST, THE HUNTER-GATHERER, WITH A WIDER VARIETY OF FOOD SOURCES, IS BETTER ABLE TO SURVIVE THE LOSS OF ONE OR MORE OF THEM.)

BEFORE THE ADVENT OF HIGH-SPEED TRANSPORT AND INTERNATIONAL RELIEF AGENCIES, EVERY LOCAL CROP FAILURE MEANT FAMINE. CHINESE OFFICIALS RECORDED 1,828 FAMINES BETWEEN THE YEARS 108 B.C. AND 1910. A BRITISH LIST RECOUNTS 11 FAMINES IN THE 13TH CENTURY ALONE.

FAMINES TEND TO BE LOCAL: A POOR HARVEST IN ONE REGION HAS LITTLE EFFECT ON FOOD SUPPLIES ELSEWHERE. THEREFORE, THE EFFECT OF FAMINE ON GLOBAL POPULATION IS MINOR, DESPITE THE DEVASTATION IN THE IMMEDIATE NEIGHBORHOOD.

MERE BLIPS IN THE CURVE!

WITH # PESTILENCE

IT'S DIFFERENT: DISEASES CAN SPREAD ACROSS CONTINENTS.

FOR EXAMPLE, IN THE 13TH CENTURY PLAGUE CAME TO EUROPE, CARRIED BY SAILORS ARRIVING FROM ASIA. BY THE END OF THE CENTURY, EUROPEAN POPULATION HAD DROPPED BY ABOUT HALF.

WELL, AT LEAST THE TAX COLLECTORS DIE, TOO...

TO UNDERSTAND EPIDEMIC DISEASES, YOU HAVE TO SEE THINGS FROM THE MICROBE'S POINT OF VIEW: THE BACTERIA, VIRUSES, AND AMOEBAS THAT CAUSE DISEASES ARE, LIKE OURSELVES, ORGANISMS LOOKING FOR A HABITAT.

TO SOME OF THESE PATHOGENS, A WARM HUMAN BODY IS LIKE HEAVEN ON EARTH.

BUT THESE LITTLE r-STRATEGISTS, BREEDING LIKE MAD, MAY *KILL THEIR HOST*, AN UNFAVORABLE OUTCOME FOR THE GERMS, WHO ALSO SUFFER A DIE-OFF.

IT'S MUCH BETTER FOR THE LITTLE INFECTIOUS AGENTS WHEN PLENTY OF HOSTS REMAIN ALIVE, AT LEAST IN SOME CONDITION.

WE'RE NOT ASKING FOR MUCH!

THEY HAVE SEVERAL STRATEGIES TO ACCOMPLISH THAT END.

THEY MAY *COEVOLVE* WITH HUMANS. PEOPLE WHO SURVIVE THE DISEASE MAY BE GENETICALLY EQUIPPED WITH A MORE EFFECTIVE IMMUNE RESPONSE TO THAT PARTICULAR MICROBE...

SOME GERMS I DON'T MIND!

THE SURVIVING PATHOGEN ALSO MUTATES TO A LESS VIRULENT FORM.

GO, DAUGHTER, AND BE A KINDER, GENTLER GERM!

THE RESULT IS A MORE RESISTANT HUMAN POPULATION, INFECTED BY A LESS DEADLY FORM OF THE DISEASE. THE DISEASE IS NOW SAID TO BE *ENDEMIC*.

HEE HEE HEE

THE SUCCESSFUL INFECTION IS ONE THAT ALLOWS ENOUGH HOSTS TO LIVE, WHILE AVOIDING DESTRUCTION BY THE HOST'S IMMUNE SYSTEM.

GOOD-BYE, GRANDMOTHER!

114

THIS EXPLAINS WHY FOREIGN DISEASES CAN BE SO DEVASTATING TO ISOLATED POPULATIONS: IT TAKES TIME TO COEVOLVE. FOR EXAMPLE, WHEN EUROPEANS FIRST LANDED IN AMERICA, THEIR DISEASES KILLED FAR MORE NATIVE AMERICANS THAN EUROPEAN WEAPONS.

NOT THAT IT MATTERS MUCH, FROM OUR POINT OF VIEW...

GETTING TO KNOW YOOU...

EUROPE, MEANWHILE, HAD ALSO FALLEN VICTIM TO HORRIBLE PLAGUES. IN 430 B.C., 160 A.D., 540, THE 1300s, ETC. ETC. ETC. USUALLY, THESE WERE IMPORTED FROM ELSEWHERE BY SAILORS.

ARH! WE'RE ABOARD THE H.M.S. "DISEASE VECTOR!"

EVENTUALLY, THE WORST PASSES, THE DISEASE BECOMES ENDEMIC, AND POPULATION AGAIN RISES STEADILY—UNTIL THE NEXT THING COMES AROUND.

WHAT'S NEXT?

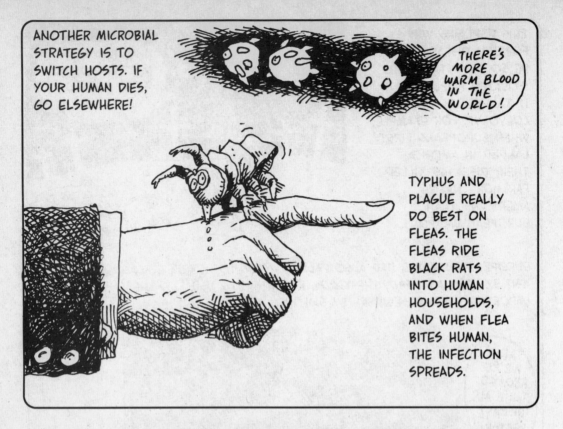

ANOTHER MICROBIAL STRATEGY IS TO SWITCH HOSTS. IF YOUR HUMAN DIES, GO ELSEWHERE!

THERE'S MORE WARM BLOOD IN THE WORLD!

TYPHUS AND PLAGUE REALLY DO BEST ON FLEAS. THE FLEAS RIDE BLACK RATS INTO HUMAN HOUSEHOLDS, AND WHEN FLEA BITES HUMAN, THE INFECTION SPREADS.

INFLUENZA PREFERS PIGS AND DUCKS. CHINA, WHERE FARMERS RAISE PIGS AND DUCKS TOGETHER, IS THE BREEDING GROUND FOR NEW STRAINS OF FLU VIRUS, WHICH CROSSES THE PACIFIC OCEAN EVERY YEAR.

WHEEZE

SNORFLE

QUACK

COUGH

CONTACT BETWEEN CULTURES SPREADS DISEASE. THUS, EPIDEMICS HAVE OFTEN OCCURRED DURING PERIODS OF PEACE, WHEN DISTANT CIVILIZATIONS COULD SAFELY MAKE CONTACT BY SAIL OR CARAVAN. SIMULTANEOUS PLAGUES IN CHINA AND ROME IN THE 160s MAY HAVE RESULTED FROM TRADE BETWEEN THE TWO COUNTRIES.

ALL ROADS LEAD TO SOME HIDEOUS ILLNESS!

BUT DISEASES ARE ALSO SPREAD BY WAR, AS ARMIES ENTER PARTS UNKNOWN. (ACCORDING TO ONE THEORY, SYPHILIS CAME TO EUROPE FROM AMERICA WITH THE CONQUISTADORS.)

WIN A FEW, LOSE A FEW...

TODAY, WHEN ANYONE WITH A PLANE TICKET CAN GO ANYWHERE, INCLUDING THE DEEPEST WILDERNESS, CONDITIONS ARE RIPE FOR A NEW PANDEMIC.

IT'S THE REVENGE OF THE RAIN FOREST!

DISEASE CAN HAVE A MAJOR IMPACT ON POPULATION LEVELS. IN ENGLAND AND FRANCE DURING THE 1200S, FOR EXAMPLE, PLAGUE CARRIED OFF ROUGHLY 1/3 OF THE POPULATION—WHILE THE WARS, REVOLTS, AND FAMINES THAT FOLLOWED IN PLAGUE'S WAKE KILLED ANOTHER 1/3.

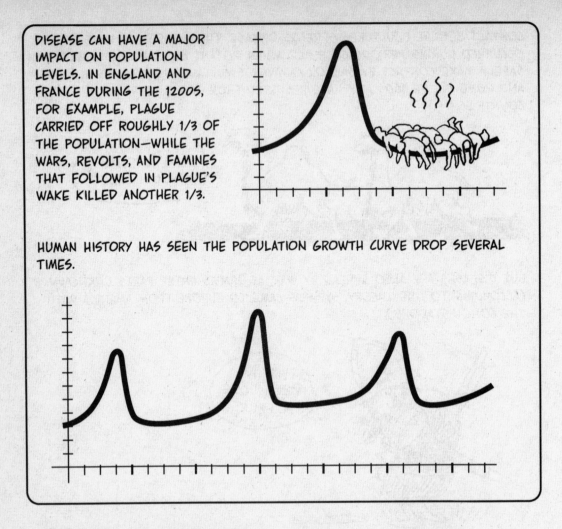

HUMAN HISTORY HAS SEEN THE POPULATION GROWTH CURVE DROP SEVERAL TIMES.

BUT WE REPRODUCE FAIRLY RAPIDLY, AND AT NO TIME WAS THE SURVIVAL OF THE SPECIES SERIOUSLY THREATENED. CERTAINLY, FOR THE PAST 500 YEARS OR SO, THE TREND HAS BEEN IN ONE DIRECTION ONLY:

WHAT DIRECTION?

IF YOU LOOK AT POPULATION IN THE WORLD TODAY, YOU GET A SURPRISE: IN THE DEVELOPED WORLD OF THE RICHER COUNTRIES, POPULATION IS BARELY GROWING, WHILE IN THE POORER, DEVELOPING COUNTRIES, IT'S SOARING.

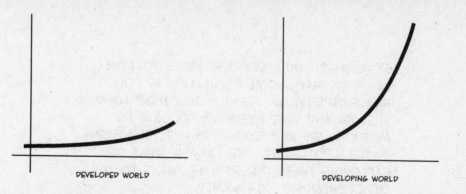

DEVELOPED WORLD

DEVELOPING WORLD

IN THE DEVELOPING WORLD, THE LIMITING FACTORS OF WAR, STARVATION, AND DISEASE STILL PLAY A ROLE, BUT THEIR EFFECTS HAVE BEEN MITIGATED BY ANTIBIOTICS AND PUBLIC SANITATION. THE HIGH BIRTH RATE OF THESE MAINLY AGRICULTURAL SOCIETIES THEN PRODUCES HIGH POPULATION GROWTH.

IT'S GOOD TO HAVE MANY CHILDREN!

ISN'T IT?

IN THE DEVELOPED WORLD, IT SEEMS, THERE IS A NEW LIMITING FACTOR FOR HUMAN POPULATION, ONE THAT MALTHUS DIDN'T FORESEE: *PROSPERITY.* WHEN PEOPLE HAVE A HIGH COMFORT LEVEL, ACCESS TO BIRTH CONTROL, AND THE SECURITY THAT THEIR CHILDREN WILL SURVIVE TO ADULTHOOD, THEY HAVE SMALL FAMILIES!!

IS IT PROSPERITY OR T.V.?

YOU ARE BECOMING A VEGETABLE... SOON YOU WILL NEED BEES TO POLLINATE YOU...

STILL, WHEN YOU ADD THE NUMBERS TOGETHER, YOU FIND THAT GLOBAL POPULATION IS STILL RISING RAPIDLY. ALL THESE PEOPLE NEED FOOD, SHELTER, AND HEAT IN THE WINTER, NOT TO MENTION CARS AND COMPUTERS AND TELEVISIONS. IN THE NEXT CHAPTER, WE EXAMINE WHAT OUR EXPLODING ENERGY NEEDS HAVE MEANT TO THE OTHER SPECIES OF THE WORLD.

• CHAPTER 9 •

BYE BYE, BIOME!

AS THE HUMAN SPECIES HAS DIVERTED BIOLOGICAL RESOURCES TO ITS OWN USE, THE EFFECT ON MOST OTHER SPECIES HAS BEEN DEVASTATING. IN THIS CHAPTER, WE SEE HOW DEFORESTATION, COMMERCIAL HUNTING, AND OTHER HUMAN ACTIVITIES HAVE AFFECTED THE BIOSPHERE... AND WE WARN YOU, IT ISN'T A PRETTY PICTURE!

HOW MANY SPECIES ARE THERE ON EARTH? NOBODY KNOWS... EVERY YEAR, SCIENTISTS DISCOVER NEW ONES, AND NO ONE CAN SAY HOW MANY MORE HAVE YET TO BE FOUND. TO DATE, SOME *1.4 MILLION* SPECIES HAVE BEEN IDENTIFIED, AND ESTIMATES FOR THE BALANCE RUN FROM ONE MILLION TO 100 MILLION(!).

MOST OF THEM ARE MOSQUITOS, I THINK!

OF THE KNOWN ANIMAL SPECIES, LESS THAN 4% ARE VERTEBRATES, AND HALF OF THESE ARE FISH. THE REMAINING 2% IS DIVIDED AMONG BIRDS (0.8%), REPTILES AND AMPHIBIANS (0.8%), AND MAMMALS (0.4%). ABOUT 85% OF ALL ANIMAL SPECIES ARE ARTHROPODS (INVERTEBRATES WITH JOINTED BODY AND LIMBS), A CLASS THAT INCLUDES INSECTS, SPIDERS, LOBSTERS, AND SCORPIONS.

HM. MAYBE INSECTS HAVE ALREADY INHERITED THE EARTH!

PERCENTAGE OF TOTAL KNOWN SPECIES

| 6% | 85% | 2% | 0.8% | 0.8% | 0.4% |
| WORMS & CLAMS & SUCH | ARTHRO-PODS | FISH | REPTILES & AMPHIBIANS | BIRDS | MAMMALS |

SPECIES ARE NOT SPREAD EVENLY AROUND THE GLOBE. SOME PARTS OF THE WORLD, MOST NOTABLY THE TROPICS, HAVE A MUCH GREATER VARIETY OF FLORA AND FAUNA THAN ELSEWHERE. HERE IS A MAP SHOWING THE REGIONS OF *MEGADIVERSITY*.

MADAGASCAR

ISLANDS IN PARTICULAR MAY BE HIGHLY DIVERSE. MADAGASCAR, FOR EXAMPLE, WHICH WAS ONCE CONNECTED TO THE AFRICAN MAINLAND, BECAME AN ISOLATED HAVEN FOR ANCIENT SPECIES THAT DIED OUT IN OTHER PLACES. TODAY, SOME 10% OF **ALL KNOWN SPECIES** LIVE IN MADAGASCAR, INCLUDING FIVE TIMES AS MANY KINDS OF TREES AS IN ALL NORTH AMERICA!

AUSTRALIA SPLIT OFF FROM ASIA WHEN MARSUPIALS WERE COMMON. IN MOST OF THE WORLD THESE PRIMITIVE MAMMALS WERE LARGELY DISPLACED BY PLACENTAL MAMMALS, BUT IN AUSTRALIA THEY FLOURISHED AND DIFFERENTIATED INTO EVERYTHING FROM KANGA-ROOS TO WOLFLIKE CARNIVORES.

I MAY LOOK LIKE A WOLF, BUT I'M REALLY A KIND OF VICIOUS POSSUM!

DESPITE THEIR DIVERSITY, ISLAND ECOSYSTEMS ARE OFTEN FRAGILE. THIS MAKES THEM A DRAMATIC MICROCOSM FOR THE EFFECTS OF HUMAN BEINGS ON THE BIOSPHERE.

HACK
HACK
HACK

ON MADAGASCAR, FOR EXAMPLE, FARMERS HAVE STRIPPED OFF OVER 90% OF THE TROPICAL FOREST, THREATENING THE MOST DIVERSE ECOSYSTEM ON EARTH.

THE SUBSEQUENT EROSION DESTROYS THE QUALITY OF FARMLAND AS WELL. (INDONESIA AND THE PHILIPPINES ARE MOSTLY CLEARED, TOO.)

WHEN EUROPEAN SHIPS BEGAN CIRCLING THE GLOBE, THEY WREAKED HAVOC ON MANY AN ISLAND.

ON ISLANDS WITHOUT LARGE PREDATORS, ANIMALS HAD NO URGE TO FLEE THE NEW HUMAN ARRIVALS. THE DODO BIRD, FOR EXAMPLE, WAS SUCH A DOCILE TARGET THAT IT WAS TOTALLY EXTINCT BY 1680.

THIS WOULD BE A LOT MORE FUN IF YOU'D MAKE A RUN FOR IT!

DITTO THE MOA, THE ELEPHANT BIRD, AND THE TASMANIAN EMU.

EUROPEAN SETTLERS CLEARCUT CARIBBEAN ISLANDS FOR SUGAR PLANTATIONS...

SORRY... BUT TREES MAKE ME NERVOUS!

IMPORTED SHEEP AND CATTLE TO AUSTRALIA, WHERE NONE HAD BEEN BEFORE. NOW THERE ARE 100 MILLION SHEEP AND 8 MILLION COWS IN AUSTRALIA.

BA·A·A·A BA·A·A·A

BAA

AND EVERYWHERE, THEY BROUGHT DOGS, PIGS, AND RATS THAT DEVOURED BIRDS' EGGS AND DISPLACED COMPETITORS.

WHAT A CRUEL JOKE! DO YOU REALIZE WHAT WE PIGS HAVE TO EAT, NOW THAT THE EGGS ARE GONE?

MORE?

MODERN COMMERCIAL "HUNTER-GATHERERS" ASSAULTED SEABIRD POPULATIONS BY HARVESTING EGGS, MEAT, AND FEATHERS FROM THEIR ISLAND NESTING GROUNDS.

AT LEAST YOU COULD FLY AWAY!

AT THE TURN OF THE CENTURY, MILLIONS OF PROCESSED ALBATROSS EGGS WENT INTO U.S. ARMY RATIONS, AND THE ALBATROSS LAYS JUST ONE EGG PER YEAR. THE ALBATROSS HAS SURVIVED, BUT THE GREAT AUK, HUNTED FOR MEAT AND FEATHERS, WENT EXTINCT BY 1844.

A MAN HAS TO EAT!

BEFORE YOU CONDEMN YOUR OWN SPECIES FOR ITS WANTON BEHAVIOR, NOTE THAT MOST OF THIS DESTRUCTION WAS DONE—AND CONTINUES TO BE DONE—TO PROVIDE PEOPLE WITH FOOD, CLOTHING, SHELTER, AND OTHER NECESSITIES.

THERE ARE VERY FEW CHICKENS IN THE MIDDLE OF THE OCEAN!

IN THE DAYS BEFORE PLASTIC AND PETROLEUM, PEOPLE USED WHALE-OIL LAMPS, FIREWOOD AND CHARCOAL, BEAVER HATS, BADGER-BRISTLE BRUSHES, DEERSKIN GLOVES, LOG HOUSES, AND WALRUS-HIDE BULLETPROOF VESTS.

GOOD THING THEY'RE NOT SHOOTING HARPOONS!

POF

ON THE OTHER HAND, IT IS A LITTLE HARDER TO JUSTIFY KILLING ELEPHANTS FOR IVORY PIANO KEYS, TORTOISES FOR FORTUNE-TELLING (IN CHINA), CHAMOIS FOR SILVER-POLISHING RAGS, OR OSTRICH, FLAMINGOS, BIRDS OF PARADISE, AND EGRETS JUST FOR THEIR ORNAMENTAL FEATHERS.

WHY NOT DISPLAY YOUR OWN DAMN SECONDARY SEX CHARACTERISTICS?

126

THE CIVILIZED SYSTEM THAT
PRODUCES THIS SLAUGHTER IS

COMMERCIAL HUNTING

(AND FISHING, FOR THAT MATTER).

UNLIKE THE SUBSISTENCE HUNTER, WHO KILLS ANIMALS FOR USE BY HIMSELF
AND HIS FAMILY, THE COMMERCIAL HUNTER SERVES A DISTANT **MARKETPLACE.**
FOR EXAMPLE, THE PLAINS INDIANS KILLED A FEW BISON AND THEN USED MEAT,
HIDE, AND ALL, WHEREAS *BUFFALO BILL* SLAUGHTERED THE ANIMALS WHOLESALE
SO BUTCHERS COULD CUT OFF THE HUMPS AND TONGUES AND SHIP THEM TO
RAILROAD WORKERS FOR DINNER, WASTING THE REST.

GENERALLY SPEAKING, THE
SUBSISTENCE HUNTER MARSHALS
HIS RESOURCES CAREFULLY. WHEN
ONE SOURCE OF FOOD RUNS LOW,
HE SWITCHES TO SOMETHING ELSE.
THIS STABILIZES THE ECOSYSTEM.

INSTABILITY
JUST MAKES
ME BARF!

COMMERCIAL HUNTING, BY CONTRAST, IS INHERENTLY UNSTABLE. WHEN AN ANIMAL RESOURCE BECOMES SCARCE, ITS PRICE GOES UP, DRIVING THE HUNTER TO HUNT EVEN MORE.

FIVE DOLLARS A FEATHER!

TEN!

FIFTEEN!

BESIDES, WHAT'S THE COMMERCIAL HUNTER TO DO? HE'S A *SPECIALIST*... HUNTING IS HIS JOB (AGAIN, UNLIKE THE HUNTER-GATHERER, WHO IS MORE OF A GENERALIST).

A BUCK'S A BUCK, AND A DUCK'S A DUCK!

NOTE: THIS IS NOT TO BE CONFUSED WITH HUNTING FOR SPORT. SPORT HUNTING HAS ALMOST ALWAYS BEEN CLOSELY REGULATED, AND SPORTSMEN LIKE THEODORE ROOSEVELT WERE AMONG THE FIRST CONSERVATIONISTS.

EXTINCTION ISN'T ANY FUN!

THE STARKEST EXAMPLES OF COMMERCIAL HUNTING'S EXCESSES COME FROM THE EUROPEAN COLONIZATION OF NORTH AMERICA, WHERE PREVIOUSLY THE INHABITANTS HAD MAINTAINED AN ENVIRONMENT OF VAST FORESTS AND ABUNDANT WILDLIFE.

WHAT A NUISANCE!

THE NUMBERS ARE STAGGERING: 379,012 POUNDS OF SPERMACETI (WHALE GLOP) EXPORTED FROM THE EAST COAST IN ONE YEAR (1770)... BETWEEN 121,355 AND 612,000 DEERSKINS SHIPPED FROM SOUTH CAROLINA EVERY YEAR BETWEEN 1706 AND 1748... 500,000 SNOWY EGRETS FROM VENEZUELA IN 1848... 130,000 BIRD SKINS IN 1892 FROM ONE FLORIDA FEATHER MERCHANT ALONE... 118,000 SEA OTTER PELTS IN 1856... FEATHERS FROM 48,000 CONDORS ON DISPLAY IN LONDON SHOPS IN 1913...

CAN YOU TAKE ANY MORE?

IN ALL, HUNTERS EXTERMINATED AN ESTIMATED TOTAL OF **60 MILLION BISON, 200 MILLION BEAVER, 5 BILLION PRAIRIE DOGS,** AND UNCOUNTED BILLIONS OF PASSENGER PIGEONS.

AND YOU THINK I'M SCARY!

TO DESTROY AN ECOSYSTEM, IT IS NOT NECESSARY TO HUNT EVERY SPECIES TO DEATH. REMOVAL OF JUST **ONE** SPECIES MAY HAVE A PROFOUND EFFECT ON OTHERS, JUST AS REMOVING THE KEYSTONE HAS AN EFFECT ON AN ARCH.

KEYSTONE

ARCH WITH KEYSTONE   .... AND WITHOUT

AND SO WE HAVE THE CONCEPT OF A

KEYSTONE SPECIES.

THESE SPECIES ARE NATURE'S **ENGINEERS,** WHOSE ACTIVITIES ALTER THE ENVIRONMENT IN WAYS THAT CREATE HABITATS FOR OTHER ANIMALS AND HELP DETERMINE THE ECOLOGY OF THE KEYSTONE SPECIES' SURROUNDINGS.

AN EXAMPLE IS THE **BEAVER,** WHOSE DAM BUILDING CREATES WETLANDS AND CONVERTS FORESTS TO MEADOWS. THE BEAVER MAKES ITS ENVIRONMENT MOISTER THAN IT WOULD HAVE BEEN, BEAVERLESS.

YOU **CAN** HARVEST TREES WITHOUT WRECKING EVERYTHING!

BUT THE BEAVER'S PELT MADE IRRESISTIBLE HATS... THE BEAVER WAS HUNTED TO NEAR EXTINCTION BY 1840... AND ALTHOUGH THE ANIMALS HAVE BOUNCED BACK SINCE THEN, THEIR CURRENT POPULATION (AROUND 6-12 MILLION) IS ONLY *5%* OF WHAT IT WAS—AND OUR LANDSCAPE HOLDS MUCH LESS WATER THAN IT USED TO.

OTHER KEYSTONE SPECIES INCLUDE THE ALLIGATOR (IT DIGS WATER-RETAINING HOLES) AND THE PRAIRIE DOG (ITS UNDERGROUND TOWNS PROVIDE HOMES FOR MANY OTHER ANIMALS, AERATE THE SOIL, AND INCREASE ITS ABSORBENCY).

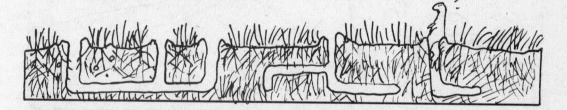

BUT RANCHERS SAW PRAIRIE DOG TOWNS AS NOTHING BUT A BUNCH OF LEG-BREAKING HOLES FOR COWS TO STEP IN. EVEN THOUGH CATTLE ACTUALLY *PREFERRED* TO DINE AMONG PRAIRIE DOGS (THE GRASS IS TENDERER THERE), RANCHERS POISONED THE PRAIRIE DOGS TO MAKE THE WORLD SAFE FOR CATTLE.

IN ADDITION TO OVERHUNTING,

COMPETITIVE EXCLUSION

HAS DEPRIVED WILD SPECIES OF THEIR HABITATS. IN THE U.S., PEOPLE HAVE PLOWED 98% OF THE GRASS-LANDS, CUT DOWN 94% OF THE VIRGIN FOREST (ALTHOUGH SOME HAS GROWN BACK), AND FILLED OVER 50% OF THE WETLANDS.

ARH! WE NEED TO EVOLVE AN ELECTRIC BEAVER!

MIGRATION ROUTES ARE INTERRUPTED BY HIGHWAYS, WHICH BLOCK THE MOVE-MENTS OF SMALL ANIMALS AS SURELY AS MOUNTAINS OR RIVERS.

EVEN WHERE PATCHES OF HABITAT ARE LEFT, THEY ARE OFTEN TOO SMALL AND DISCONNECTED TO SUSTAIN A VIABLE POPULATION: THE "ISLAND EFFECT."

TODAY, IN AN EFFORT TO PRESERVE AS MUCH BIO-DIVERSITY AS POSSIBLE, ECOLOGISTS TRY TO IDENTIFY PLANTS AND ANIMALS AT RISK OF EXTINCTION WHILE THEY'RE STILL ALIVE.

IT ISN'T EASY, 'CAUSE THEY CAN BE HARD TO FIND!

AN **ENDANGERED SPECIES** IS ONE WITH SO FEW SURVIVORS THAT THE SPECIES IS AT IMMEDIATE RISK OF EXTINCTION OVER ALL OF ITS RANGE. EXAMPLES INCLUDE THE CALIFORNIA CONDOR (FEWER THAN 100 KNOWN INDIVIDUALS), THE JAVAN TIGER, AND THE AFRICAN WHITE RHINOCEROS.

OVER 100 LARGE VERTEBRATE SPECIES HAVE FEWER THAN 100 INDIVIDUALS LEFT...

IN SHORT: HUNTED, HUNTED, HABITAT DESTRUCTION, HABITAT DESTRUCTION, HABITAT DESTRUCTION, HUNTED, HUNTED, HUNTED.

AN ENDANGERED SPECIES MAY HAVE ONE OR MORE OF THESE CHARACTERISTICS: LARGE SIZE, (EASILY HUNTED), FEW OFFSPRING (SLOW TO REPRODUCE), UNIQUE HABITAT (NOWHERE ELSE TO GO), SPECIALIZED DIET (KILL THEIR FOOD AND THEY'RE DEAD), TOP OF THE FOOD CHAIN (POISONED BY POLLUTION, MORE WAYS TO DISRUPT THEIR FOOD SUPPLY), EAT LIVESTOCK (SO THEY'RE HUNTED), MIGRATORY (EASILY HUNTED), OR OTHER RISKY BEHAVIOR (DITTO).

THREATENED SPECIES

ARE THOSE WITH REASONABLE POPULATIONS IN THEIR NATURAL HABITATS, BUT WHOSE NUMBERS ARE FALLING OR HABITATS ARE SHRINKING. EXAMPLES INCLUDE THE BALD EAGLE AND THE GRIZZLY BEAR, WHICH WOULD BOTH BE ENDANGERED, IF NOT FOR CURRRENT ENVIRONMENTAL LAWS.

WHEN IT COMES TO ASSESSING THE HEALTH OF AN ECOSYSTEM, WE RELY ON

INDICATOR SPECIES.

THESE ARE THE FIRST TO BE AFFECTED WHEN THEIR ENVIRONMENT IS BEING DEGRADED.

IN THE U.S., MIGRATORY SONGBIRD POPULATIONS HAVE FALLEN BY HALF. THEIR NORTHEASTERN HOMES HAVE BEEN FRAGMENTED, CREATING EDGE HABITATS FOR PREDATORS LIKE COWBIRDS, JAYS, RACCOONS, AND HOUSE CATS, WHILE THEIR WINTER QUARTERS, THE TROPICAL RAINFORESTS, ARE BEING LOGGED.

AMPHIBIANS ARE ALSO INDICATOR SPECIES, SINCE THEIR THIN SKIN AND PERMEABLE EGGS EXPOSE THEM TO POLLUTANTS IN AIR, WATER, AND SOIL. AMPHIBIAN POPULATIONS HAVE BEEN FALLING WORLDWIDE, EVEN IN HABITATS THAT APPEAR SUPERFICIALLY HEALTHY.

WELL, HOW BAD IS IT?

IT'S HARD TO ESTIMATE THE RATE OF SPECIES EXTINCTION HISTORICALLY, BUT IT HAS DEFINITELY ACCELERATED IN RECENT YEARS. BY THE 1970s, 1000 SPECIES A YEAR WERE GOING EXTINCT(!), AND BY 1990 THE ANNUAL RATE HAS SOARED TO AN ASTONISHING *4000-6000*, ACCORDING TO HARVARD BIOLOGIST E.O. WILSON.

VIRTUALLY ALL FLORA AND FAUNA ARE DECLINING, AND THOUSANDS OF UNKNOWN SPECIES ARE ALMOST CERTAINLY FALLING WITH THE TROPICAL RAINFORESTS.

IN 1989, ECOLOGISTS AT STANFORD UNIVERSITY CALCULATED AN ESTIMATE OF HOW MUCH OF THE EARTH'S NPP (NET PRIMARY PRODUCTION—I.E., USABLE PLANT BIOMASS) IS NOW USED BY HUMANS.

THEY CONCLUDED THAT WE NOW CONTROL AROUND

39%

OF ALL LAND-BASED NPP. WE ACTUALLY EAT ONLY 3%, BUT WE CONSUME THE OTHER 36% IN THE FORM OF CROP WASTES, FOREST CLEARING, DESERTIFICATION, AND SETTLEMENT.

UM... WHAT'S OUR DOUBLING TIME AGAIN?

HUMAN POPULATION IS EXPECTED TO DOUBLE IN THE NEXT 50 YEARS. UNLESS WE CHANGE OUR PATTERNS OF USAGE AND CONSUMPTION, WE WILL BE USING NEARLY ALL LAND-BASED NPP AND MUCH OF THE OCEAN'S, AND WORLD BIODIVERSITY WILL BE A THING OF THE PAST.

WHY SHOULD WE CARE IF GLOBAL DIVERSITY DISAPPEARS? BECAUSE, IN EFFECT, WE ARE PERFORMING AN IMMENSE EXPERIMENT ON THE PLANET, TAKING SYSTEMS APART AND THROWING AWAY THE PIECES. WE KNOW VERY LITTLE ABOUT THE NATURAL WORLD: HOW ANIMALS LEARN, NAVIGATE, OR COMMUNICATE, WHAT MEDICINAL PLANTS STILL LURK IN THE DEPTHS OF THE RAIN FOREST—AND MOST CRITICALLY, WHETHER THE SIMPLIFIED ECOSYSTEM WE ARE SO BUSILY CREATING CAN SUSTAIN ITSELF.

~ CHAPTER 10 ~

ENERGY WEBS

HUMAN BEINGS, ALONE AMONG ANIMALS, HAVE FOUND WAYS TO USE **NON-FOOD ENERGY** IN OUR LIVES. BURNING WOOD AND OIL HEAT OUR HOMES; ELECTRICITY POWERS OUR ELECTRIC TOOTHBRUSHES, LIGHTS OUR BULBS, AND RUNS OUR COMPUTERS; GASOLINE FIRES OUR CARS. WE SQUANDER THE STUFF, ESPECIALLY PETROLEUM PRODUCTS, AS IF THERE WERE NO TOMORROW.

WHAT IS ENERGY, ANYWAY? (WE'VE MANAGED TO GO 9 CHAPTERS WITHOUT SAYING EXACTLY.) FOR THE MOMENT, LET'S PRETEND THERE ARE ONLY TWO KINDS OF ENERGY, OR AT LEAST TWO WAYS TO THINK ABOUT ENERGY: AS **HEAT** AND **WORK**.

heat
and
work

ON THE ONE HAND, ENERGY CAN BE THOUGHT OF AS AN AMOUNT OF **HEAT.** IF YOU BURN FUEL, THE ENERGY RELEASED IS (NEARLY) THE SAME AS THE TOTAL AMOUNT OF HEAT GIVEN OFF. (SOME ENERGY IS ALSO EMITTED AS LIGHT.)

ON THE OTHER HAND, IN MECHANICAL TERMS, ENERGY IS THE SAME AS **WORK.** IF YOU PUSH SOMETHING WITH A FORCE OVER A DISTANCE, THE ENERGY EXERTED IS (NEARLY) THE PRODUCT OF THE FORCE TIMES THE DISTANCE. (SOME ENERGY IS ALSO DISSIPATED AS HEAT DUE TO FRICTION.)

Force

← DISTANCE →

Energy = Force × distance

FOR MOST OF HUMAN HISTORY, THESE WERE THE KINDS OF ENERGY PEOPLE NEEDED: HEAT AND WORK.

O.K. SOMEBODY GO WORK!

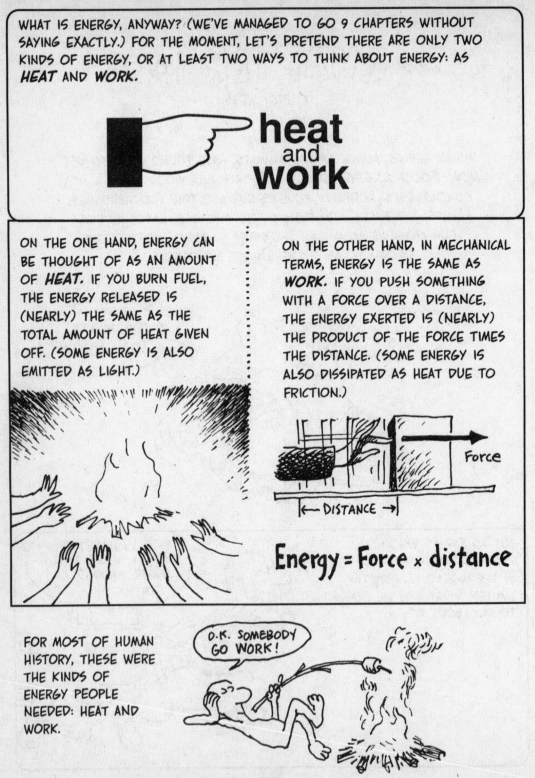

AT FIRST, THERE WAS BASICALLY ONLY ONE SOURCE OF HEAT, AND THAT WAS *BIOMASS:* WOOD, STRAW, OR, WHERE WOOD WAS SCARCE, DRIED COW DUNG (AS IN INDIA), ETC.

I DESERVE TO BE WORSHIPPED.

MECHANICAL ENERGY ALSO CAME FROM BIOMASS: ANIMAL POWER--OR BRUTE FORCE--AND THE BRUTE IN QUESTION USUALLY HAD TWO LEGS.

WHO CAN AFFORD FOUR LEGS?

EVEN WHEN ANIMALS WERE HARNESSED, THEIR FUEL EFFICIENCY WAS LOWER THAN HUMANS': I.E., THEY CONSUMED MORE FOOD TO DO THE SAME WORK. ECONOMICS DICTATED THAT HUMANS DO MOST OF THE WORK, SAVING ANIMALS FOR THE REALLY HEAVY LIFTING AND HAULING.

GEE-YAP!

UH. OH!

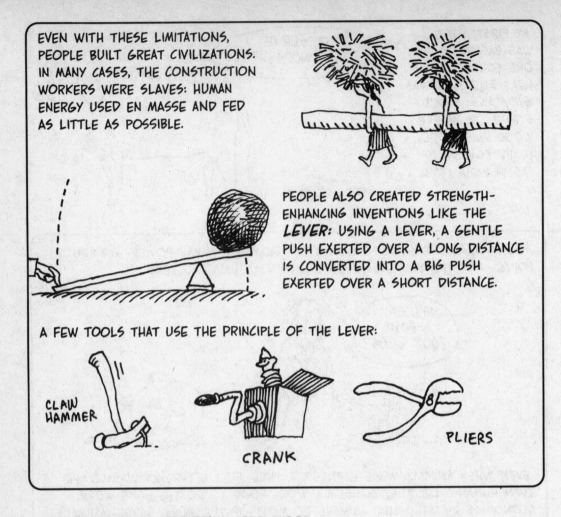

EVEN WITH THESE LIMITATIONS, PEOPLE BUILT GREAT CIVILIZATIONS. IN MANY CASES, THE CONSTRUCTION WORKERS WERE SLAVES: HUMAN ENERGY USED EN MASSE AND FED AS LITTLE AS POSSIBLE.

PEOPLE ALSO CREATED STRENGTH-ENHANCING INVENTIONS LIKE THE *LEVER:* USING A LEVER, A GENTLE PUSH EXERTED OVER A LONG DISTANCE IS CONVERTED INTO A BIG PUSH EXERTED OVER A SHORT DISTANCE.

A FEW TOOLS THAT USE THE PRINCIPLE OF THE LEVER:

CLAW HAMMER

CRANK

PLIERS

HEAT-ENHANCERS WERE FOUND TOO: WOOD WAS REDUCED TO CHARCOAL, WHICH BURNS HOTTER, AND OVENS, KILNS, AND SMELTERS WERE DESIGNED TO BAKE ANYTHING FROM BREAD TO BRICK TO METAL. WHENEVER POSSIBLE, THESE FACTORIES WERE BUILT NEAR FORESTS, THE FUEL SUPPLY.

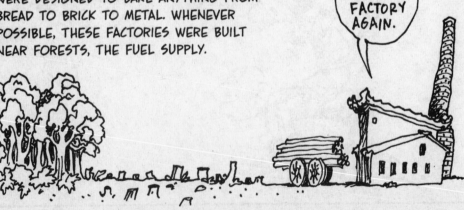

TIME TO MOVE THE FACTORY AGAIN.

(AND LET'S NOT FORGET THE MACHINERY DRIVEN BY WATER WHEELS AND WIND POWER.)

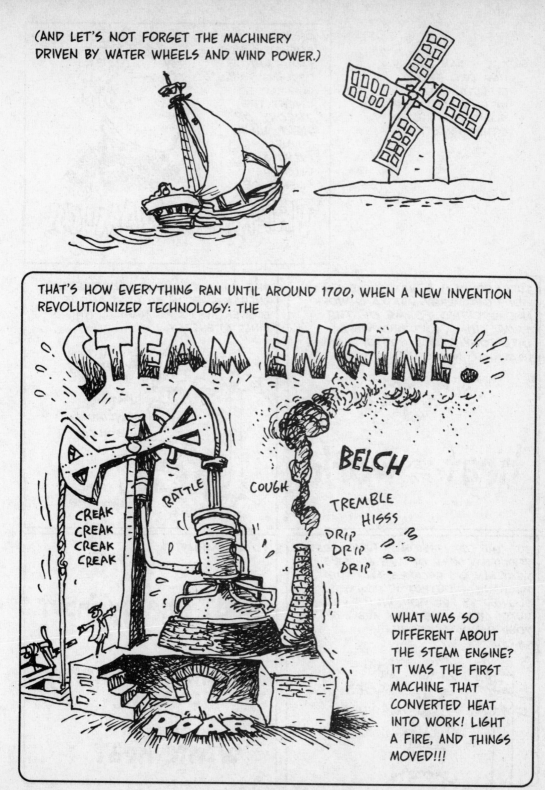

THAT'S HOW EVERYTHING RAN UNTIL AROUND 1700, WHEN A NEW INVENTION REVOLUTIONIZED TECHNOLOGY: THE

STEAM ENGINE.

CREAK CREAK CREAK CREAK

RATTLE

COUGH

BELCH

TREMBLE HISSS

DRIP DRIP DRIP

ROAR

WHAT WAS SO DIFFERENT ABOUT THE STEAM ENGINE? IT WAS THE FIRST MACHINE THAT CONVERTED HEAT INTO WORK! LIGHT A FIRE, AND THINGS MOVED!!!

THE STEAM ENGINE HAD TWO GREAT EFFECTS, ONE INTELLECTUAL AND SCIENTIFIC, THE OTHER PRACTICAL.

ON THE SCIENTIFIC SIDE, SADI **CARNOT** (1796-1832) WAS INSPIRED TO INVENT THE **THEORY OF HEAT,** WHICH TURNED EVENTUALLY INTO THE SCIENCE OF

THERMODYNAMICS.

THERMODYNAMICS UNITES THE TWO VIEWS OF ENERGY: HEAT AND WORK ARE JUST **TWO FORMS OF THE SAME THING,** AND THEY WERE **INTERCONVERTIBLE.** YOU CAN CHANGE HEAT INTO MOTION, AND VICE VERSA.

VOILA!

heat ⟷ work

THE STEAM ENGINE IS A **CONVERTOR:** ENERGY GOES IN AS HEAT AND IS CONVERTED TO MECHANICAL (OR KINETIC) ENERGY.

MOTION ENERGY OUT

PISTON
CYLINDER
BOILER

FURNACE

HEAT ENERGY IN

BUT THE CONVERSION IS NEVER 100% EFFICIENT. SOME OF THE BOILER'S HEAT ALWAYS ESCAPES, AND THE MECHANICAL OUTPUT IS ALWAYS SLOWED BY FRICTION, WHICH PRODUCES HEAT. (JUST TRY RUBBING YOUR HANDS TOGETHER.)

SQUEE
SQUEE
SQUEE

TO SUMMARIZE,

heat energy input
⇩
mechanical energy output
➕
waste heat

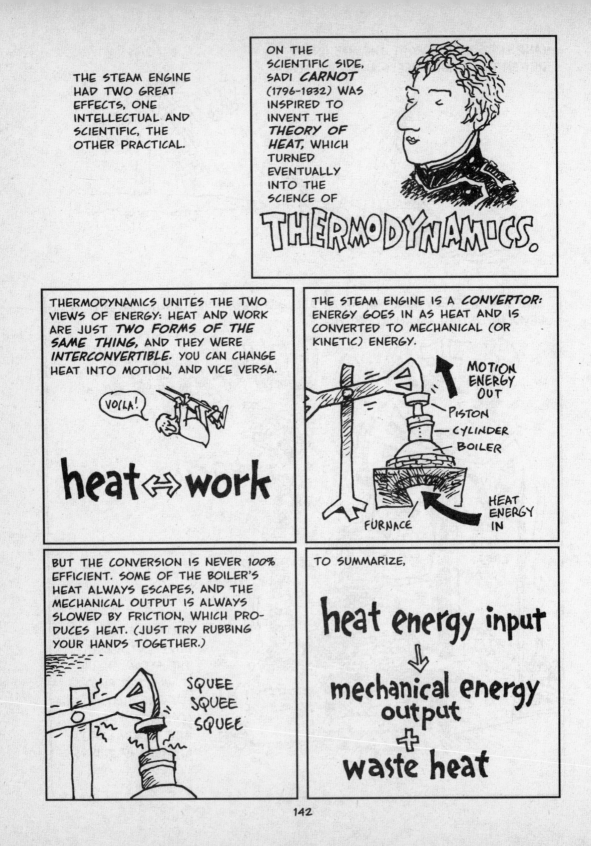

IT WAS SOON RECOGNIZED THAT ALL ENERGY, NOT JUST HEAT AND WORK, WERE REALLY THE SAME. GRAVITY, ELECTRICITY, CHEMICAL REACTIONS, SOUND, LIGHT: ALL WERE TYPES OF ENERGY THAT MIGHT BE CONVERTED, CHAMELEONLIKE, ONE TO THE OTHER.

CLAUSIUS

FOR EXAMPLE, *GRAVITATIONAL ENERGY* MAKES WATER FLOW DOWNSTREAM, TURNING A WATER WHEEL, WHICH CAN DO WORK, WHETHER GRINDING GRAIN OR HOISTING MINERALS FROM A MINE.

AGAIN, THE MILL IS A **CONVERTOR**, CHANGING GRAVITATIONAL ENERGY INTO MECHANICAL ENERGY.

OH, THAT GRAVITY!

ALSO, BY THE *SECOND LAW OF THERMODYNAMICS* (SEE P. 70), EVERY MACHINE MUST **WASTE SOME ENERGY**, MAINLY DISSIPATING IT INTO THE ENVIRONMENT AS HEAT.

IT'S DEPRESSING, REALLY!

"SQUEE" SQUEE SQUEE

BOLTZMANN

IN OTHER WORDS, ALL ENERGY TECHNOLOGY CONSISTS OF **CONVERTORS**, WHICH ARE ALWAYS LESS THAN 100% EFFICIENT.

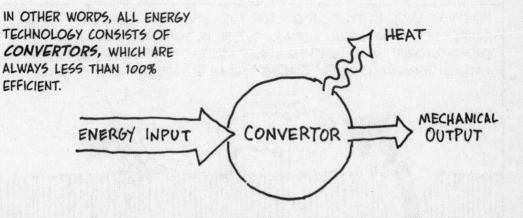

ENERGY INPUT → CONVERTOR → MECHANICAL OUTPUT

HEAT

THE SECOND, MORE PRACTICAL EFFECT OF THE STEAM ENGINE WAS AN INCREASE IN THE DEMAND FOR STUFF TO *BURN.*

FEED ME!

ALREADY BY 1700, ENGLAND WAS MOSTLY DEFORESTED BY WOODCUTTERS SEEKING FUEL AND FARMERS CLEARING FIELDS. WHERE WAS FUEL TO COME FROM?

HAVE YOU CONSIDERED COW DUNG?

WE'RE PROUD AND BRITISH! IT WOULD HAVE TO BE SHEEP DUNG!

THE ANSWER TURNED OUT TO BE THIS UNIMPRESSIVE-LOOKING STUFF:

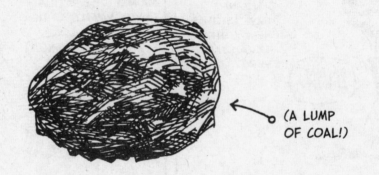

(A LUMP OF COAL!)

HERE WAS ANOTHER REVOLUTION: FOR THE FIRST TIME, PEOPLE DREW HEAT ENERGY NOT FROM LIVING BIOMASS, BUT FROM A *FOSSIL FUEL:* SOLAR ENERGY STORED IN AN IMMENSE, BURIED RESERVOIR OF LONG-DEAD PLANT MATERIAL. GRADUALLY, *NEW ENERGY* BEGAN ENTERING THE BIOSPHERE...

THE STEAM ENGINE
WAS FOLLOWED BY AN
EXPLOSION, AND NOT
ONLY OF BOILERS...
BUT ALSO AN
EXPLOSION OF
INVENTIONS AND
DISCOVERIES.

WE HAVE **GOT** TO FIND A WAY TO SCREW THESE THINGS TOGETHER!

OVER THE NEXT 200 YEARS CAME A SERIES OF NEW CONVERTORS, NEW TECHNOLOGIES, NEW FUELS, EVEN NEW FORMS OF ENERGY, LIKE ELECTRICITY.

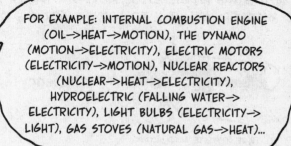

FOR EXAMPLE: INTERNAL COMBUSTION ENGINE (OIL->HEAT->MOTION), THE DYNAMO (MOTION->ELECTRICITY), ELECTRIC MOTORS (ELECTRICITY->MOTION), NUCLEAR REACTORS (NUCLEAR->HEAT->ELECTRICITY), HYDROELECTRIC (FALLING WATER-> ELECTRICITY), LIGHT BULBS (ELECTRICITY-> LIGHT), GAS STOVES (NATURAL GAS->HEAT)...

JUST LIKE THE INVENTION OF AGRICULTURE, THE USE OF FOSSIL FUELS PUT NEW ENERGY IN HUMAN HANDS. THE EFFECTS WERE SIMILAR: **MORE PEOPLE** AND **MORE ORGANIZATION.** THIS TIME IT WAS CALLED THE

INDUSTRIAL REVOLUTION.

REMEMBER: IT ALL RUNS ON CONTINUOUS ENERGY INJECTIONS FROM OUTSIDE THE SYSTEM!

AN ENERGY NETWORK RESEMBLES A FOOD WEB. START WITH AN ENERGY SOURCE, LIKE AN OIL WELL... THE FUEL MUST BE EXTRACTED, USING UP SOME ENERGY IN THE PROCESS (TO RUN THE PUMP, FOR EXAMPLE)... TRANSPORTED, AGAIN AT SOME ENERGY COST... STORED... DISTRIBUTED... AND CONVERTED TO USEFUL FORM. EACH STAGE WASTES SOME ENERGY AND LOWERS THE OVERALL EFFICIENCY OF THE SYSTEM.

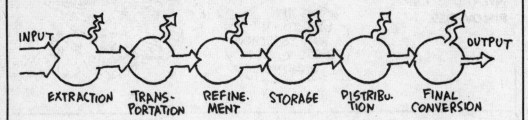

JUST AS WITH FOOD WEBS, THE EFFICIENCY AT EACH STAGE IS THE PERCENTAGE OF ENERGY THAT SURVIVES THE PROCESS IN USABLE FORM. FOR EXAMPLE, TO COMPUTE THE EFFICIENCY OF MOVING A TANKER FULL OF OIL FROM ONE PLACE TO ANOTHER, YOU HAVE TO ACCOUNT FOR THE OIL USED IN DRIVING THE TRUCK.

$$\text{efficiency} = \frac{\text{oil in tanker}}{\text{oil in tanker} + \text{oil used in driving it around}}$$

TO BE COMPLETE, ALSO INCLUDE SOME OF THE ENERGY USED TO MANUFACTURE AND MAINTAIN THE TRUCK, AND THE HIGHWAY, TOO!

THE OVERALL EFFICIENCY AT ANY STAGE OF THE ENERGY NETWORK IS THE PRODUCT OF THE EFFICIENCIES OF ALL THE PREVIOUS STAGES. HERE IS A HYPOTHETICAL EXAMPLE:

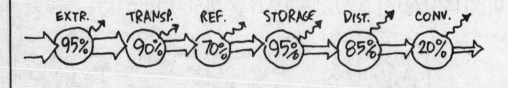

$$0.95 \times 0.9 \times 0.7 \times 0.95 \times 0.85 \times 0.2 = 0.097 \sim \mathbf{10\%}$$

IF YOU THINK THIS LOOKS INCREDIBLY WASTEFUL, YOU ARE RIGHT! A TYPICAL AUTOMOBILE ENGINE IS AT MOST ABOUT 20% EFFICIENT AS A CONVERTOR. IF THE SYSTEM THAT EXTRACTS, REFINES, AND DISTRIBUTES PETROLEUM PRODUCTS IS 50% EFFICIENT (A REASONABLE ESTIMATE), THEN THE ENERGY ACTUALLY DELIVERED TO A CAR'S WHEELS IS NO MORE THAN *10%* OF THE CHEMICAL ENERGY IN THE ORIGINAL PETROLEUM.

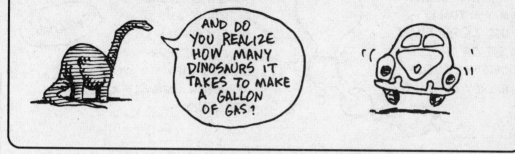

AND DO YOU REALIZE HOW MANY DINOSAURS IT TAKES TO MAKE A GALLON OF GAS?

BUT WHEN IT COMES TO INEFFICIENCY, THE KING OF CONSUMPTION, THE SULTAN OF SLACK, THE WAZIR OF WASTE, IS **ELECTRICITY.**

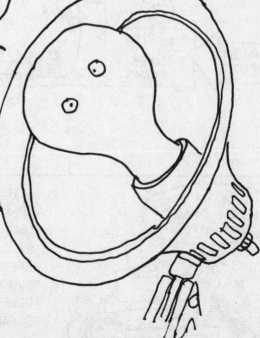

MAYBE I'M NOT AS BRIGHT AS I OUGHT TO BE...

ONE REASON IS SIMPLE: MANY ELECTRICAL CONVERTORS ARE GROSSLY INEFFICIENT. AN INCANDESCENT LIGHT BULB, FOR EXAMPLE, HAS AN EFFICIENCY OF AROUND 5%. *95% OF THE ELECTRIC ENERGY USED BY A LIGHT BULB IS WASTED AS HEAT.*

THAT 5% COMES AFTER ALL THE LOSSES IN EXTRACTION, TRANSPORT, ETC., SO AN ELECTRIC LIGHT MAY ACTUALLY USE **LESS THAN 2%** OF THE ORIGINAL FUEL'S ENERGY CONTENT.

I'M SORRY!

THIS IS TRUE ONLY OF INCANDESCENT BULBS. FLUORESCENT LIGHTS ARE SOMEWHAT MORE EFFICIENT, AND SOME EXPERIMENTAL SULFUR-BASED LAMPS APPROACH 100% EFFICIENCY.

SHOCKING!

ELECTRIC POWER ALSO WASTES ENERGY BY INSERTING ANOTHER LEVEL OF CONVERSION IN THE ENERGY NETWORK. INSTEAD OF CONVERTING FOSSIL FUEL DIRECTLY TO USEFUL FORM (MOTION, HEAT, LIGHT, ETC.), AN ELECTRIC POWER PLANT FIRST CONVERTS FUEL TO ELECTRICITY, WHICH THEN FLOWS THROUGH WIRES (LOSING SOME ENERGY IN THE PROCESS!) TO ELECTRICAL APPLIANCES, THE ULTIMATE CONVERTORS FOR USE.

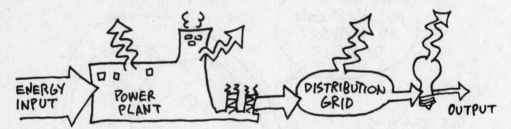

ENERGY INPUT — POWER PLANT — DISTRIBUTION GRID — OUTPUT

I WANT A HORSE-DRAWN COMPUTER...

TO ITS CREDIT, ELECTRICITY IS FLEXIBLE, VERSATILE, EASILY DISTRIBUTED, AND CLEAN—AT LEAST, AT THE CONSUMER LEVEL, IT'S CLEAN. IT CAN ALSO BE GENERATED BY WATER OR WIND POWER, REDUCING OUR DEPENDENCE ON FOSSIL FUELS, AND FOR SOME USES, IT'S HARD TO IMAGINE AN ALTERNATIVE.

FOR MOST OF THE CENTURY BETWEEN 1870 AND 1970, ENERGY SUPPLIES HAVE BEEN RELATIVELY CHEAP AND ABUNDANT, ESPECIALLY IN THE UNITED STATES, WHICH HAD VAST RESERVES OF OIL AND COAL.

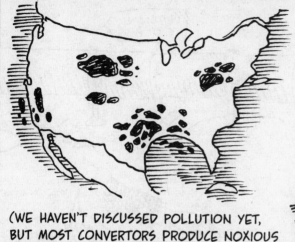

UNDER THESE CONDITIONS, THE U.S.A.'S NATIONAL ENERGY STRATEGY WAS SIMPLE: EXTRACT IT FAST, USE LOTS OF IT, AND DON'T WORRY TOO MUCH ABOUT EFFICIENCY OR POLLUTION.

(WE HAVEN'T DISCUSSED POLLUTION YET, BUT MOST CONVERTORS PRODUCE NOXIOUS GASES, ASH, OR OTHER COMBUSTION BY-PRODUCTS. MORE ON THIS LATER!)

WORLD ENERGY CONSUMPTION ROSE *SIXTY-FOLD* BETWEEN 1860 AND 1985— ESPECIALLY IN THE U.S.A, WHERE CHEAP GAS CREATED A CAR CULTURE...

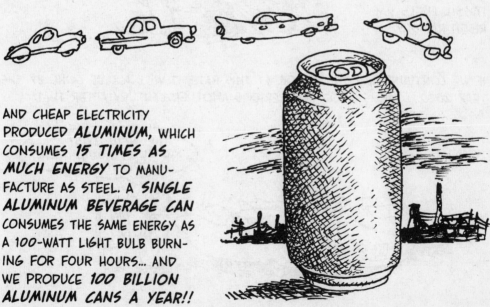

AND CHEAP ELECTRICITY PRODUCED *ALUMINUM,* WHICH CONSUMES *15 TIMES AS MUCH ENERGY* TO MANU-FACTURE AS STEEL. A *SINGLE ALUMINUM BEVERAGE CAN* CONSUMES THE SAME ENERGY AS A 100-WATT LIGHT BULB BURN-ING FOR FOUR HOURS... AND WE PRODUCE *100 BILLION ALUMINUM CANS A YEAR!!*

WESTERN EUROPE HAD LESS OIL AND COAL THAN NORTH AMERICA, BUT COUNTRIES LIKE BRITAIN, FRANCE, AND GERMANY HAD THE FINANCIAL AND MILITARY POWER TO GAIN ACCESS TO OVERSEAS ENERGY RESERVES.

WITH ITS MORE EXPENSIVE OIL, EUROPE HAS MANAGED TO BUILD A HIGHLY INDUSTRIAL ECONOMY WITH AROUND HALF THE PER-CAPITA ENERGY CONSUMPTION OF THE UNITED STATES.

AS THE WORLD EMBRACES THE JOYS OF INDUSTRIALIZATION, THE DEMAND FOR FOSSIL FUELS HAS RISEN RAPIDLY.

IF WE CONTINUE TO CONSUME OIL AT THIS RATE, IT WILL ALL BE GONE BY THE YEAR 2030. COAL IS GOOD FOR PERHAPS ANOTHER CENTURY AFTER THAT.

WOULD YOU CALL THIS AN

energy crisis?

WHY NOT?

FINDING SOLUTIONS TO THE ENERGY PROBLEM IS COMPLICATED BY THE FACT THAT ENERGY CONSUMPTION IS SO UNEQUAL. THE UNITED STATES, WITH 4% OF THE WORLD'S POPULATION, USES ABOUT **25%** OF ALL THE ENERGY. IN FACT, ONE ESTIMATE SAYS WE USE 10% OF THE GLOBAL ENERGY BUDGET JUST DRIVING TO WORK AND BACK. OBVIOUSLY, WE HAVE SOME WASTE TO CUT!

BUT IN MUCH OF THE WORLD, PEOPLE ARE POOR AND HAVE LITTLE ACCESS TO ENERGY. THE AVERAGE CITIZEN OF INDIA, FOR EXAMPLE, USES AROUND **2-3%** THE ENERGY CONSUMED BY THE AVERAGE AMERICAN. FOR THE 2 BILLION PEOPLE WHO DEPEND ON FIREWOOD FOR FUEL, THE ENERGY CRISIS LOOKS ENTIRELY DIFFERENT. THEY WANT MORE HIGH-QUALITY ENERGY.

151

WHATEVER THE OVERALL SOLUTION TO THE ENERGY CRISIS MAY BE, IT HAS TO INCLUDE STEPS BY THE INDUSTRIALIZED, HIGH-CONSUMING NATIONS TO REDUCE THEIR FOSSIL FUEL USAGE. WHERE CAN THE SAVINGS COME FROM: LET'S SIMPLIFY THE DIAGRAM OF AN ENERGY SYSTEM, LUMPING ALL THE CONVERSION AND TRANSPORTATION STEPS INTO ONE:

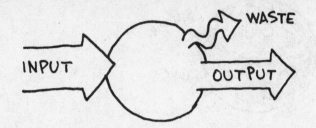

THERE ARE ONLY THREE POSSIBLE AREAS OF POTENTIAL SAVINGS, AREN'T THERE? THERE'S THE INPUT, THE WASTE, AND THE OUTPUT!

output:

THIS INVOLVES MAKING CHOICES THAT USE LESS ENERGY: BUYING BEVERAGES IN BOTTLES INSTEAD OF CANS... WALKING, TAKING PUBLIC TRANSIT, OR CAR-POOLING INSTEAD OF DRIVING YOUR OWN CAR... TURNING THE THERMOSTAT DOWN IN WINTER, ETC...

THESE ARE PERSONAL, INDIVIDUAL CHOICES, BUT AS ENERGY PRICES RISE THEY TURN INTO SOCIAL POLICY AND BUSINESS DECISIONS: DESIGNING COMMUNITIES WITH HOUSING CLUSTERS NEAR BUSINESS AND SHOPS, SUBSIDIES FOR MASS TRANSIT, DAYCARE IN THE WORKPLACE TO REDUCE PARENTS' DRIVING, ETC.

waste:

REDUCING WASTE MEANS INCREASING THE SYSTEM'S EFFICIENCY.

SOURCE REDUCTION

MEANS AVOIDING WASTE BY USING ALTERNATIVE PRODUCTS: CLOTH SHOPPING BAGS INSTEAD OF PAPER, TEA INSTEAD OF CANNED DRINKS, ETC.

RECYCLING DIVERTS WASTE ENERGY BACK INTO THE SYSTEM AS INPUT. RECYCLING AN ALUMINUM CAN MEANS THE ENERGY OF MANUFACTURE ISN'T CONSUMED IN ONE USE.

THERE ARE MANY, MANY OTHER WAYS TO INCREASE ENERGY EFFICIENCY, SUCH AS MORE EFFICIENT PRODUCTS, LIKE HIGH-MILEAGE AUTOS OR BETTER-INSULATED HOMES, OR REFRIGERATORS WITH FANS TO COOL THE COILS, OR FURNACES THAT CAPTURE CHIMNEY HEAT, OR BETTER LUBRICANTS TO REDUCE FRICTION.

WE'RE MAKING PROGRESS HERE!

THERE ARE ALSO INEFFICIENCIES OF THE WHOLE SYSTEM: TOO MANY CONVERSION STEPS, BADLY DESIGNED TRANSPORTATION SYSTEMS, ETC. THESE ARE HARD PROBLEMS TO SOLVE, AS IN SAN FRANCISCO, A CITY WHICH HAS NEVER BEEN ABLE TO BRING MASS TRANSIT ALL THE WAY TO THE AIRPORT!

ALL IT TAKES IS THE WILL, THE MONEY, AND A FEW THOUSAND MEETINGS...

input:

YOU MIGHT SAY THE INPUT WILL TAKE CARE OF ITSELF. WHEN THE OIL RUNS OUT, ENERGY USE IS BOUND TO GO DOWN!

LOOK ON THE BRIGHT SIDE: NOBODY STEALS CARS ANYMORE!

TO MAKE THIS TRANSITION FROM PETROLEUM USE AS SMOOTH AS POSSIBLE, IT'S NECESSARY TO DO WHAT HUMANITY HAS HISTORICALLY DONE: CONSIDER ALTERNATIVE ENERGY SOURCES.

AT ONE TIME,

NUCLEAR POWER

LOOKED LIKE "THE" ANSWER. IN A NUCLEAR REACTOR, RADIOACTIVE FUEL RODS GLOW HOT, DRIVING A HIGH-TECH STEAM ENGINE THAT CAN GENERATE ELECTRICITY, TURN A SHIP'S SCREW, ETC. BECAUSE IT BURNS NOTHING, A NUCLEAR REACTOR EMITS NO SOOT OR OTHER POLLUTING COMBUSTION BY-PRODUCTS.

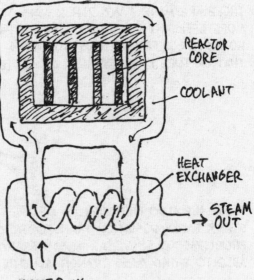

REACTOR CORE

COOLANT

HEAT EXCHANGER

STEAM OUT

WATER IN

OH, IT'S SO LOVELY! LOVELY!

THERE IS ALSO THE PROMISE OF A "BREEDER REACTOR," WHICH CAN ACTUALLY PRODUCE MORE FUEL THAN IT CONSUMES. ENERGY FOREVER!

BUT OH, THOSE NEGATIVES! RADIATION IS BAD FOR YOUR HEALTH, SO SPECIAL PRECAUTIONS MUST BE TAKEN IN MINING, PROCESSING, AND TRANSPORTING FUEL. IT CAN ALSO BE USED FOR BOMBS, SO YOU HAVE TO GUARD IT! EVERYTHING ABOUT NUCLEAR ENERGY IS VERY EXPENSIVE.

OOPS!

THEN THERE'S THE QUESTION OF WHAT TO DO WITH THE SPENT FUEL, WHICH STAYS "HOT" BASICALLY FOREVER.

WORST OF ALL, SYSTEM FAILURES CAN REALLY RUIN THE NEIGHBORHOOD...

FOR THE NEXT 10,000 YEARS.

AS I WAS SAYING, "OOPS!"

OOPS AGAIN!

WORLDWIDE, THERE ARE CURRENTLY ABOUT 520 NUCLEAR POWER PLANTS EITHER WORKING OR UNDER CONSTRUCTION, PRODUCING SOME 5% OF THE WORLD'S TOTAL ELECTRICITY... BUT BEYOND THESE, FEW NEW ONES ARE EXPECTED.

SOLAR POWER

FOSSIL FUEL'S ENERGY ULTIMATELY COMES FROM THE SUN. WHY NOT GO DIRECTLY TO THE SOURCE AND ELIMINATE SOME CONVERTORS?

PHOTOVOLTAIC CELLS (SOLAR PANELS) CONVERT SUNLIGHT DIRECTLY TO ELECTRICITY. HOWEVER, THEY ARE STILL EXPENSIVE AND INEFFICIENT, MAKING LARGE-SCALE ELECTRICAL GENERATION UNLIKELY FOR SOME TIME. SOLAR HEATING AND HOT WATER ARE FEASIBLE NOW (IN SUNNY PLACES).

HEY! THAT'S MY SUNLIGHT!

HYDROELECTRIC

PLANTS PRODUCE IMMENSE AMOUNTS OF ELECTRICITY, USING THE CLEANEST, CHEAPEST "FUEL" OF ALL: FALLING WATER.

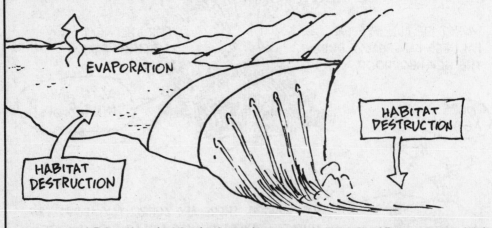

EVAPORATION

HABITAT DESTRUCTION

HABITAT DESTRUCTION

THE DOWNSIDE: THE LARGE DAMS REQUIRED BY HYDROELECTRIC POWER HAVE ENVIRONMENTAL AND SOCIAL COSTS: BEHIND THE DAM, WATER IS LOST THROUGH EVAPORATION, AND FERTILE SILT SETTLES TO THE BOTTOM OF THE LAKE. DOWNSTREAM, REDUCED WATER FLOW DRIES OUT THE ENTIRE WATERSHED.

WIND

CAN BE USED EITHER TO GENERATE ELECTRICITY OR TO DO WORK DIRECTLY. IN SOME AREAS, WINDMILLS CAN GENERATE ENOUGH POWER TO LIGHT CITIES. BUT WINDMILLS BREAK OFTEN, AND ARE HARD ON BIRDS.

BIOMASS—

WOOD, FOR EXAMPLE—IS STILL A VIABLE FUEL. IT'S RENEWABLE, AND CAN BE CONSERVED BY THE USE OF EFFICIENT STOVES. PLANTING FAST-GROWING TREE SPECIES AMELIORATES DEFORESTATION. BUT BURNING BIOMASS IS POLLUTING.

GEOTHERMAL

ENERGY TAPS THE HEAT OF THE EARTH'S INTERIOR TO MAKE ELECTRICITY, HEAT WATER, OR DO OTHER USEFUL WORK. ALL YOU NEED IS A DEEP HOLE AND SOME CONVERSION TECHNOLOGY. THE HEAT IS DOWN THERE, AND IT'S FREE!

AND THE DOWNSIDE?

I DON'T KNOW... WE HAVEN'T HIT BOTTOM YET!

TO SUM UP, THERE REALLY IS AN ENERGY PROBLEM: MODERN LIFE DEPENDS ON LARGE, DAILY INJECTIONS OF ENERGY FROM— **SOMEWHERE.** AT PRESENT, IT'S FOSSIL FUELS, BUT THEY CAN'T LAST FOREVER.

SIGH...

A TECHNOLOGICAL SOLUTION TO THIS PROBLEM, IF IT EXISTS, WILL PROBABLY HAVE TO INCLUDE MANY INGREDIENTS: LOWER-CONSUMPTION LIFESTYLES IN THE DEVELOPED WORLD, MORE EFFICIENT CONVERTORS, AND SAFE, NON-POLLUTING ALTERNATIVE SOURCES OF ENERGY. THERE ARE INNUMERABLE SPECIFIC AREAS TO WORK ON, FROM TRANSPORTATION TO MORE EFFICIENT COOKSTOVES FOR THE THIRD WORLD...

HEY! AND WHY NOT A SOLAR-POWERED TOOTHBRUSH?

WILL THE EFFORT PAY OFF?

NOT IF WE DON'T MAKE IT!!

· CHAPTER 11 ·

LET'S EAT AGAIN!

IN THE LAST CHAPTER, WE DESCRIBED ENERGY NETWORKS AS IF THEY WERE
SEPARATE FROM FOOD WEBS. IN FACT, THE TWO KINDS OF ENERGY ARE
CLOSELY RELATED. IN THE FIRST PLACE, FOOD IS AT THE BASE OF THE
WHOLE ENERGY CHAIN. IF NOBODY EATS, NOBODY PUMPS GAS!

SECOND, MOST FOOD IS GROWN USING CHEMICAL FERTILIZERS, WHICH ARE
MADE FROM PETROLEUM PRODUCTS. THAT IS, FOOD ENERGY IS DERIVED FROM
FOSSIL FUELS!

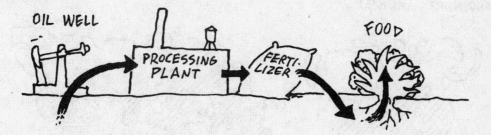

FINALLY, THE GLOBAL INDUSTRIAL ECONOMY PLAYS A MAJOR ROLE IN DECIDING
WHICH CROPS ARE PLANTED AND WHERE THEY ARE SHIPPED.

AGRICULTURAL PRACTICES VARY WIDELY FROM COUNTRY TO COUNTRY AND
REGION TO REGION. IN THE INDUSTRIALIZED WEST, WE USE FOSSIL-FUEL
DEVOURING HEAVY FARM MACHINERY AND PLENTY OF FERTILIZERS AND
PESTICIDES, WHILE IN ENERGY-POOR REGIONS, FARMERS RELY ON MUSCLE
POWER AND COOK WITH WOOD OR CHARCOAL.

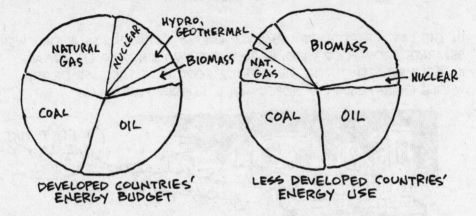

DEVELOPED COUNTRIES'
ENERGY BUDGET

LESS DEVELOPED COUNTRIES'
ENERGY USE

AT PRESENT, ABOUT 25% OF THE ARABLE LAND IS USED FOR **NOMADIC
HERDING**... ANOTHER 25% IS UNDER OLD-FASHIONED **SHIFTING CULTIVATION**,
AND THE REST IS DIVIDED BETWEEN **TRADITIONAL INTENSIVE AGRICULTURE**
(MOSTLY IN ASIA) AND **LARGE-SCALE INDUSTRIAL AGRICULTURE** (EUROPE
AND NORTH AMERICA).

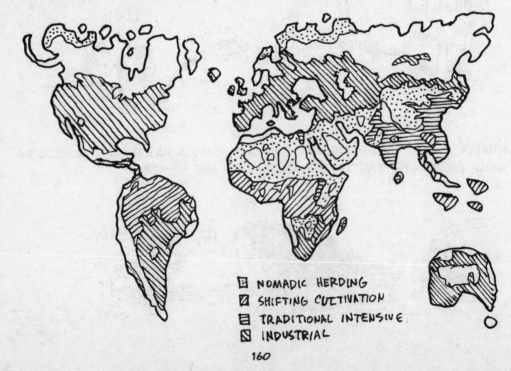

NOMADIC HERDING
SHIFTING CULTIVATION
TRADITIONAL INTENSIVE
INDUSTRIAL

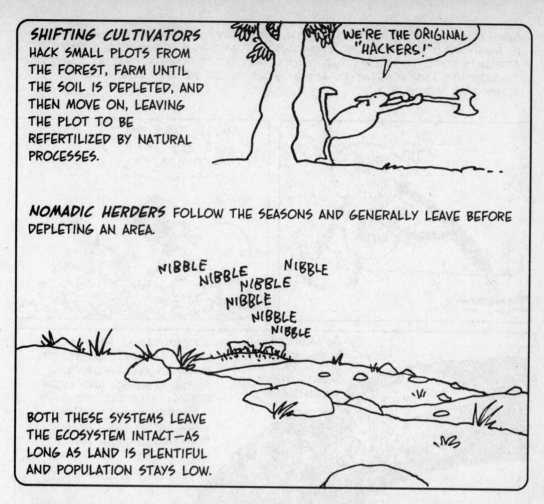

SHIFTING CULTIVATORS HACK SMALL PLOTS FROM THE FOREST, FARM UNTIL THE SOIL IS DEPLETED, AND THEN MOVE ON, LEAVING THE PLOT TO BE REFERTILIZED BY NATURAL PROCESSES.

WE'RE THE ORIGINAL "HACKERS!"

NOMADIC HERDERS FOLLOW THE SEASONS AND GENERALLY LEAVE BEFORE DEPLETING AN AREA.

NIBBLE NIBBLE NIBBLE NIBBLE NIBBLE NIBBLE NIBBLE NIBBLE

BOTH THESE SYSTEMS LEAVE THE ECOSYSTEM INTACT—AS LONG AS LAND IS PLENTIFUL AND POPULATION STAYS LOW.

PIGS WILL "RECYCLE" ANYTHING!

OO!

TRADITIONAL INTENSIVE FARMING PRODUCES VERY HIGH CROP YIELDS BY USING LOTS OF LABOR AND RECYCLING NUTRIENTS EFFICIENTLY. CHINESE FARMERS, WHO CYCLE NUTRIENTS THROUGH PIGS, CHICKENS, DUCKS, FISH, AND ALGAE, RETURNING ORGANIC MATTER TO THE SOIL, ACHIEVED OUTPUT ROUGHLY **TEN TIMES** THAT OF PREINDUSTRIAL EUROPE. CHINA HAS MAINTAINED REASONABLY GOOD SOIL FERTILITY FOR MILLENNIA.

INDUSTRIAL AGRICULTURE, WHICH RELIES ON HEAVY MACHINERY AND CHEMICALS, IS **HIGH-INPUT AGRICULTURE**. IT DRAWS ENERGY FROM FOSSIL FUELS AND CONVERTS IT TO FOOD OR OTHER CROPS. (BUT DON'T FORGET THAT AT HARVEST TIME, FRUITS AND VEGETABLES ARE MOSTLY PICKED BY VERY LOW-PAID TEMPORARY WORKERS.)

EQUIPMENT, FUEL, ETC.

FERTILIZER

OIL.

WELL, THEY DON'T GIVE **US** THAT MUCH INPUT!

TO MARKET, TO MARKET, TO BUY A FAT HOG FUTURES CONTRACT!

AN INDUSTRIAL FARM NEEDS LARGE AMOUNTS OF CAPITAL (A FANCY NAME FOR MONEY), AMPLE ACREAGE, AND CHEAP ENERGY. LIKE ALL INDUSTRY, INDUSTRIAL AGRICULTURE GREW UP IN THE ERA OF PLENTIFUL FOSSIL FUEL.

I CONVERT OIL INTO CHOLESTEROL!

WHEN FUEL IS CHEAP, THERE IS LITTLE INCENTIVE TO CONSERVE IT, SO INDUSTRIAL FARMING HAS BEEN ALMOST UNBELIEVABLY WASTEFUL. GRAIN-FED CATTLE, FOR EXAMPLE, CONSUME AROUND **7 TIMES** AS MUCH ENERGY AS THEY PRODUCE AS BEEF.

162

DESPITE ITS LOWER EFFICIENCY, HIGH-INPUT AGRICULTURE TENDS TO PUSH OUT MORE TRADITIONAL, MORE ENERGY-EFFICIENT METHODS.

THE MARKETPLACE HAS ITS OWN SPECIAL KIND OF EFFICIENCY!

* * * * * * *
* AND IN FACT,
* THE ENTIRE
* FOSSIL-BASED
* GLOBAL ECONOMY
* HAS A PRO-
* FOUND EFFECT
* ON AGRICULTURE
* EVERYWHERE.

THE REASON IS NOT HARD TO FIND: THE PEOPLE WHO CONTROL *ENERGY* ULTIMATELY CONTROL *WEALTH*.

THE RICHEST COUNTRIES— AND THE RICHEST COMPANIES— CAN CONTROL *MARKETS*, *TECHNOLOGY*, *MINERAL RESOURCES*, AND *LAND USE*.

WE CAN COMPARE NATIONAL WEALTH ON THE BASIS OF *GROSS NATIONAL PRODUCT*, OR *GNP*. THIS IS THE CASH VALUE OF ALL GOODS AND SERVICES PRODUCED EACH YEAR. IN 1993, FOR EXAMPLE, THE USA HAD A GNP OF "ONLY"

$5 TRILLION.

(NOT CHICKEN FEED!)

WELL, SOME OF IT WAS...

BECAUSE SOME COUNTRIES ARE MORE POPULOUS THAN OTHERS, IT HELPS TO LOOK AT *GNP PER CAPITA* OR GNP DIVIDED BY POPULATION. THIS INDICATES THE AVERAGE PRODUCTION PER PERSON. THE RICHEST 10 COUNTRIES' GNP PER CAPITA IS AROUND *$20,000*, WHILE THE POOREST 40 COUNTRIES COME IN AT *LESS THAN $300*.

THIS GRAPH SHOWS HOW WEALTH IS SPREAD AROUND THE WORLD. THE HEIGHT OF EACH BAR REPRESENTS THE NUMBER OF COUNTRIES WITH A GIVEN GNP PER CAPITA.

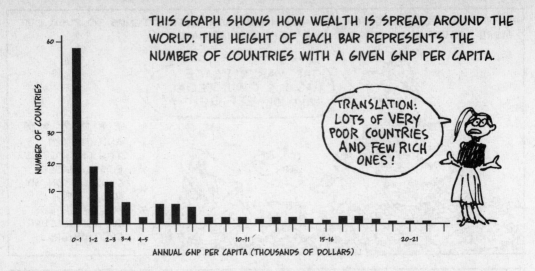

TRANSLATION: LOTS OF VERY POOR COUNTRIES AND FEW RICH ONES!

BECAUSE INDUSTRIALIZATION HAS BEEN THE KEY TO WEALTH, NEARLY EVERY COUNTRY ON EARTH WANTS *MORE INDUSTRY.* BUT HOW TO GET IT? THE BUSINESSES OF THE WEST AREN'T GIVING IT AWAY!

WE'RE NOT IN THE CHARITY BUSINESS!

WELL, DUH!

THE USUAL WAY IS FOR POOR COUNTRIES TO SELL *AGRICULTURAL PRODUCTS* ON THE WORLD MARKET, AND USE THE MONEY TO BUY OIL AND MACHINERY.

YOU SEE, THERE IS A WAY!

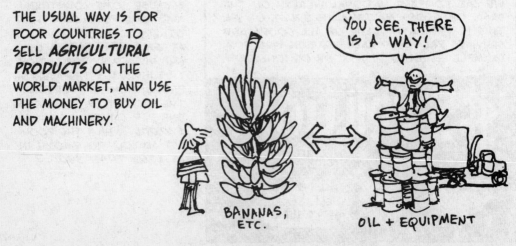

BANANAS, ETC.

OIL + EQUIPMENT

BUT THE WORLD MARKET ISN'T ESPECIALLY INTERESTED IN LOCAL FOOD CROPS LIKE MANIOC, TAPIOCA, YAMS, OR BREADFRUIT. THE WORLD WANTS CHOCOLATE, COFFEE, BANANAS, CASHEWS, AND PISTACHIO NUTS, OR NON-FOOD PLANT PRODUCTS LIKE COPRA, JUTE, AND RUBBER...

WELL, I CAN'T SELL THIS... THIS... WHAT **IS** THIS?

RESULT: LANDOWNERS PLANT THESE *"CASH CROPS"* ON LAND THAT MIGHT HAVE BEEN GROWING FOOD INSTEAD.

IS IT MY FAULT THE WORLD WANTS BANANAS?

THE WORLD HAS **GONE** BANANAS!

SO... A COUNTRY LIKE *GHANA*, WITH A PER CAPITA GNP OF *$400*, PLANTS HALF ITS FARMLAND IN *COCOA*... OR *MEXICO*, WHERE THE PRICE OF CORN KEEPS RISING AND MILLIONS GO HUNGRY, GROWS *MANGOES*, *TOMATOES*, AND *MELONS* FOR CALIFORNIA. EVEN *INDIA* EXPORTS *RICE!*

OTHER POOR FOOD EXPORTERS: PHILIPPINES (PINEAPPLE), HONDURAS, GUATE-MALA (BANANAS), UGANDA (COFFEE)...

SCIENCE TO THE RESCUE!

IN THE 1960s AND 1970s, AGRICULTURAL SCIENTISTS TRIED TO ADDRESS THE PROBLEM OF WORLD HUNGER BY CREATING THE **GREEN REVOLUTION.**

(IN THOSE DAYS, EVERYTHING WAS A REVOLUTION!)

THEY BRED VERY HIGH-YIELD VARIETIES OF WHEAT, RICE, AND MILLET, WITH HEAVY HEADS OF GRAIN, SHORT, TOUGH STEMS TO SUPPORT THEM, AND A SHORT GROWING SEASON (ALLOWING DOUBLE CROPPING). THESE TYPES COULD RAISE CROP YIELDS BY *TWO TO THREE TIMES.*

MIRACULOUS! SOMETHING FOR NOTHING!

MANY POOR COUNTRIES EMBRACED THE GREEN REVOLUTION... AND THEN MARKET FORCES TOOK OVER. IT TURNED OUT THERE WAS A DOWNSIDE TO THE NEW, REVOLUTIONARY GREENS: THEY NEEDED A LOT OF FERTILIZER AND WATER. THEIR HIGH OUTPUT DEPENDED ON HIGH INPUT!

WHO CAN AFFORD INPUT?

ONLY A FEW AFFLUENT FARMERS COULD AFFORD THE FERTILIZER, ETC. THEY PLANTED THE NEW VARIETIES... THEIR LAND PRODUCED MORE... GRAIN PRICES DROPPED... AND THEIR POOR NEIGHBORS, STUCK WITH THE OLD VARIETIES, BECAME EVEN POORER.

HERE! LESS THAN YOU CAN LIVE ON!

RICE

THIS MAKES ME SEE **RED**!

DISCLAIMER: THE OPINIONS EXPRESSED BY CHARACTERS IN THIS BOOK ARE NOT NECESSARILY THOSE OF THE AUTHORS.

THE SUCCESSFUL ONES BOUGHT MORE LAND, ENLARGING THEIR HOLDINGS AT THE EXPENSE OF THEIR NEIGHBORS. LANDLESS FARMERS FLOCKED TO THE CITIES, HOPING FOR JOBS—IN AN INDUSTRIAL ECONOMY THAT BARELY EXISTED YET.

RUMBLE

I WANT ONE OF THEM JOHN DEERE CAPS!

RESULT: MORE INDUSTRIAL AGRICULTURE, MORE FOOD, AND MORE POVERTY!

WHAT GOOD IS FOOD IF YOU DON'T EAT IT?

IN INDIA, 75% OF FARMERS ARE POORER NOW THAN THEY WERE IN 1975, EVEN THOUGH INDIA PRODUCES MORE FOOD PER PERSON NOW THAN IT DID THEN.

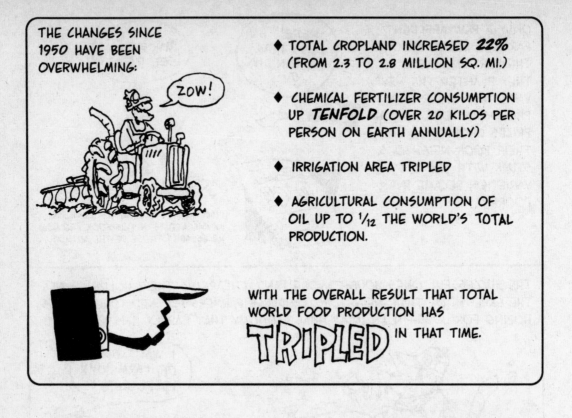

THE CHANGES SINCE 1950 HAVE BEEN OVERWHELMING:

ZOW!

◆ TOTAL CROPLAND INCREASED *22%* (FROM 2.3 TO 2.8 MILLION SQ. MI.)

◆ CHEMICAL FERTILIZER CONSUMPTION UP *TENFOLD* (OVER 20 KILOS PER PERSON ON EARTH ANNUALLY)

◆ IRRIGATION AREA TRIPLED

◆ AGRICULTURAL CONSUMPTION OF OIL UP TO $\frac{1}{12}$ THE WORLD'S TOTAL PRODUCTION.

WITH THE OVERALL RESULT THAT TOTAL WORLD FOOD PRODUCTION HAS **TRIPLED** IN THAT TIME.

SINCE POPULATION HAS "ONLY" DOUBLED IN THE SAME PERIOD, THERE IS MORE FOOD PER PERSON NOW THAN AT ANY TIME IN THE PAST 50 YEARS...

ISN'T THAT SOME KIND OF PROGRESS?

BUT WE'VE SEEN THE COSTS: ALTHOUGH THERE'S MORE FOOD FOR EVERYONE TO EAT, "EVERYONE" ISN'T EATING IT. INSTEAD, WE HAVE GROWING MIDDLE AND UPPER CLASSES, WHOSE STANDARD OF LIVING HAS RISEN SHARPLY, WHILE MOST OF THE POPULATION HAS LESS.

SOMEBODY IS ALWAYS COMPLAINING ABOUT SOMETHING!

MEANWHILE, TRADITIONAL AGRICULTURE IS LOSING GROUND, AND THE SYSTEM DEPENDS MORE AND MORE ON HIGH INPUTS. CAN THIS BE SUSTAINED?

ASIDE FROM THE LIMITS ON WORLD PETROLEUM RESERVES, THE **PHOSPHORUS** CYCLE ALSO LIMITS INDUSTRIAL AGRICULTURE. CHEMICAL FERTILIZERS USE PHOSPHATES DUG FROM MINES, WHICH SHOULD PLAY OUT AROUND THE YEAR 2050 AT PRESENT RATES OF CONSUMPTION.

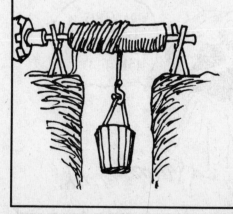

(TRADITIONAL INTENSIVE AGRICULTURE GETS ITS PHOSPHATES FROM FISH, BONE MEAL, AND RECYCLED WASTE.)

THESE CHANGES TAKE THEIR TOLL ON THE NATURAL ECOSYSTEMS AND THEIR DIVERSITY.

EVERYTHING WE SAID IN CHAPTER 7 ABOUT FARMING'S EFFECTS ON WILD SYSTEMS GOES DOUBLE FOR INDUSTRIAL AGRICULTURE.

IT SIMPLIFIES THE ECOSYSTEM!

INDUSTRIAL GROWERS USUALLY RAISE JUST ONE OR TWO VARIETIES OF WHEAT OR CORN, APPLES OR PEACHES, INSTEAD OF THE HUNDREDS OF LOCAL VARIETIES THAT USED TO THRIVE.

LACK OF VARIETY INCREASES VULNERABILITY TO PESTS AND DISEASES!

THE UBIQUITOUS RED DELICIOUS APPLE

MEANWHILE, MOST OF THE WORLD'S POTENTIAL FARMLAND LIES IN THE TROPICAL RAIN-FOREST, WITH ITS COMPLEX ECOLOGY AND POOR SOIL. THE RESULT IS AN IRRETRIEVABLE LOSS OF BIODIVERSITY, AND AN EROSION OR EXHAUSTION OF TOPSOIL.

WOA...THIS IS GETTING DEPRESSING, ISN'T IT?

AND THEN THERE'S THE ISSUE OF MEAT.

MEEEE?

A COW OR A SHEEP OR A PIG IS A **CONVERTOR**, CHANGING GRAIN OR GRASS ENERGY INTO MEAT. EACH ANIMAL REPRESENTS MORE INEFFICIENCY IN THE FOOD WEB.

IN OTHER WORDS, A PLOT OF LAND THAT PRODUCES *10,000* CALORIES WORTH OF GRAIN CAN PRODUCE AT MOST *1,000* CALORIES OF BEEF.

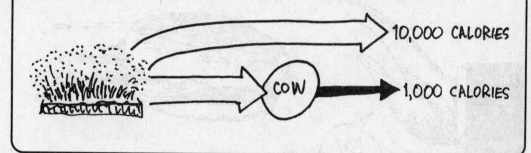

10,000 CALORIES

COW

1,000 CALORIES

AS PEOPLE GROW MORE AFFLUENT, THEY EAT MORE MEAT. AT PRESENT, TOTAL WORLD PRODUCTION OF GRAIN STANDS AT 2150 MILLION TONS, BUT ABOUT *40%* GOES INTO (NONHUMAN) ANIMAL MOUTHS. ANOTHER MAJOR CONVERTOR OF GRAIN IS *BEER...*

IF EVERYONE IN CHINA DRANK 2 BEERS A DAY, WORLD GRAIN CONSUMPTION WOULD RISE *10%* !!

AS WE'VE SEEN, (P. 135), HUMANS TAKE ALMOST 40% OF THE LAND'S NET PRIMARY PRODUCTION, BUT ONLY 3% IS ACTUAL CROP. THE REST IS FOREST CLEARING, SETTLEMENT BUILDI NG, CHAFF, AND WASTE.

AS POPULATION RISES TOWARD 6, 7, 8 BILLION, HUMANITY WILL SEEK WAYS TO WRING MORE FOOD FROM THE EARTH, AND ONE WAY IS SURELY TO IMPROVE EFFICIENCY. IF WE COULD RAISE THAT 3% TO 6%, THERE WOULD BE PLENTY OF FOOD.

FOR INSTANCE, IF MEAT-EATING WENT DOWN, A LEVEL OF CONVERTORS WOULD BE REMOVED...

HOW IS THAT TO BE DONE? BY DOUBLE-CROPPING, CAREFUL LAND USE, MORE LABOR-INTENSIVE CONSERVATION PRACTICES ON INDUSTRIAL FARMS, LESS SPOILAGE, AND *FEWER HAMBURGERS.*

JUST SAY "NO"!

BIOTECHNOLOGY ALSO HOLDS OUT SOME HOPE, FOR EXAMPLE, IN PRODUCING DISEASE-RESISTANT PLANT VARIETIES, AND EVEN IMPROVING PLANT AND ANIMAL CONVERSION EFFICIENCY.

NOW IF WE CAN JUST MAKE BURGERS GROW ON TREES!

ANOTHER APPROACH IS LOCAL

EMPOWERMENT.

IT'S HARD—IMPOSSIBLE—TO SOLVE THESE PROBLEMS FROM AFAR. ACTION IS MORE EFFECTIVE WHEN TAKEN BY THE PEOPLE DIRECTLY ON THE SPOT.

BUT THEY NEED HELP.

A CURRENT EFFORT INVOLVES SPREADING THE USE OF MORE EFFICIENT *COOKING STOVES*, WHICH PRODUCE THE SAME HEAT FROM MUCH LESS WOOD.

THE RATIONALE: IF POOR RURAL AFRICAN WOMEN DIDN'T HAVE TO SPEND 6-8 HOURS A DAY COLLECTING WOOD, THEY MIGHT HAVE MORE TIME TO DO OTHER WORK—AND DEFORESTATION WOULD SLOW, AS WELL.

WHAT DOES THIS HAVE TO DO WITH AGRICULTURE? *FORESTS CAN BE AN ESSENTIAL PART OF THE AGRICULTURAL ENERGY WEB.* THEY CATCH MOISTURE AND GENERATE NUTRIENTS THAT TRICKLE DOWNSTREAM.

AND FINALLY, WE NEED TO SAY A WORD ABOUT FISH. SEAFOOD HAS ALWAYS BEEN A TASTY, CONVENIENT SOURCE OF ANIMAL PROTEIN, AND IT MIGHT SOUND LIKE A GOOD ALTERNATIVE TO BEEF, PORK, OR LAMB.

UNFORTUNATELY, FISHING HAS ALSO TURNED INDUSTRIAL, AND DEMAND IS HUGE. TO FEED A HUMAN POPULATION IN THE BILLIONS, FISHING FLEETS HAVE NOW EXHAUSTED MANY OF THE WORLD'S PRIME FISHING GROUNDS. FISH POPULATIONS, ESPECIALLY THE PREFERRED KINDS, ARE FALLING, SO HUNTERS ARE TURNING TO SMALLER FRY AND SPECIES THAT WERE PREVIOUSLY SCORNED.

USES 15 CALORIES OF FUEL TO ACQUIRE 1 CALORIE OF FISH.

THE COMMITTEES THAT "REGULATE" FISHING HAVE USUALLY BEEN DOMINATED BY THE FISHING INDUSTRY... AND GOVERNMENTS ARE INCREASINGLY DOMINATED BY NON-FARMING CITY DWELLERS, AS CITIES SWELL WITH INDUSTRIALIZATION. IN OUR NEXT CHAPTER, WE LOOK AT URBAN ECOLOGY...

MEANWHILE, LET'S FINISH WHAT'S ON OUR PLATES...

174

 CHAPTER 12

BRIGHT LIGHTS, BIG CITY

ENERGY CONSUMPTION HAS FUELED INDUSTRY... ENERGY CONSUMPTION HAS SWELLED POPULATION... ENERGY CONSUMPTION HAS CHANGED AGRICULTURE... AND ENERGY CONSUMPTION HAS INCREASED ORGANIZATION...

IN OTHER WORDS, PEOPLE HAVE BEEN MOVING OFF THE FARM AND INTO THE CITIES...

CITIES PLAY MANY ROLES IN THE ECOSPHERE... WE CAN SUMMARIZE THEM ALL BY SAYING THAT A CITY IS WHERE **SURPLUSES OF ENERGY AND MATERIAL ARE CONCENTRATED AND TRANSFORMED.**

ON THE ONE HAND, CITIES ARE MARKETPLACES, MANUFACTURING CENTERS, AND DISTRIBUTION HUBS. GOODS FLOW IN... FACTORIES HUM... PEOPLE BUY AND SELL... AND GOODS FLOW OUT...

CITIES ARE ALSO CENTERS OF **ORGANIZATION.** ENERGY FLOWS IN TO FEED, HEAT, HOUSE, ENTERTAIN, AND MAKE CONNECTIONS BETWEEN PEOPLE WHO MOVE INFORMATION AROUND. THESE INCLUDE ADMINISTRATORS AND MANAGERS, BOTH CORPORATE AND CIVIC, PLUS EVERYONE INVOLVED IN GUIDING THOUGHT AND OPINION, FROM RELIGIOUS LEADERS TO THE MEDIA.

IN OTHER WORDS, CITY-DWELLERS MAKE DECISIONS THAT INFLUENCE THE FLOW OF ENERGY THROUGHOUT THE REST OF THE ECOSPHERE.

TO SUSTAIN THE FLOW OF THINGS AND PEOPLE, TO BRING IN FOOD FOR MILLIONS, TO DISTRIBUTE IT TO COUNTLESS SHOPS AND RESTAURANTS, A CITY MUST HAVE A *TRANSPORTATION SYSTEM.*

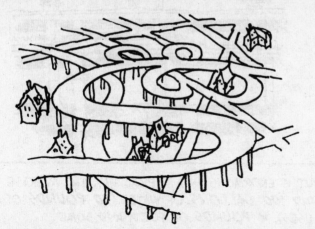

RAILROADS, RAIL YARDS, AIRPORTS, STREETS, HIGHWAYS, PARKING LOTS... IN SOME CITIES, LIKE LOS ANGELES, THE TRANSPORTATION SYSTEM COVERS MORE THAN HALF THE LAND AREA!

WITHIN THE CITY, PEOPLE USE THREE TYPES OF TRANSPORTATION:

INDIVIDUAL TRANSIT: CAR, MOTORCYCLE, MOPED, WALKING, OR BICYCLE.

NOTE: BIKES ARE FASTER AND MORE EFFICIENT THAN WALKING!

MASS TRANSIT: BUSES, TROLLEY, TRAINS.

AND **PARA** TRANSIT: CAR POOL, VAN POOL, AND JITNEY.

AIRPORT SHUTTLE

IN THE U.S., MASS TRANSIT ACCOUNTS FOR ONLY 7% OF ALL PASSENGER TRAVEL, COMPARED WITH 15% IN THE FORMER WEST GERMANY, AND ALMOST 50% IN JAPAN. THE AMERICAN WAY IS STILL ONE CAR, ONE PERSON.

ISN'T FREEDOM WONDERFUL?

BESIDES THE TRANSPORTATION NETWORK, THERE HAVE TO BE OTHER SYSTEMS FOR DELIVERING SUCH THINGS AS **WATER, ELECTRICITY, NATURAL GAS,** ETC. A CITY NEEDS PIPELINES, AQUEDUCTS, RESERVOIRS, POWER PLANTS, AND WIRES.

THE CITY'S APPETITE FOR INPUT IS ENORMOUS. IN A TYPICAL DAY, THE AVERAGE U.S. CITY DWELLER USES AROUND **100 GALLONS** OF WATER, **50 POUNDS** OF FUEL (INCLUDING INDUSTRIAL USES), **4 POUNDS** OF FOOD, AND SOME CONSUMER GOODS.

BRINGING IN GOODS AND ENERGY IS EASIER THAN GETTING RID OF THE OUTPUT: PER PERSON, THAT'S **4.5 POUNDS OF GARBAGE, A POUND OF HUMAN WASTE, 100 GALLONS OF WASTE WATER,** AND **A POUND OF AIR POLLUTION** PER DAY.

WHERE IS ALL THIS STUFF SUPPOSED TO GO? WE'VE OUTGROWN THE DAYS WHEN PEOPLE COULD JUST THROW SLOPS OUT THE WINDOW FOR PIGS TO EAT.

HISTORY HAS BEEN MOSTLY DOWNHILL FOR THE URBAN PIG...

FOR DISPOSAL PURPOSES, WASTE CAN BE DIVIDED INTO TWO KINDS: **SOLID** AND **LIQUID** (OR WASTEWATER).

AND THAT HEAVENLY KIND IN BETWEEN!

LET'S START WITH THE SOLID—OTHERWISE KNOWN AS GARBAGE.

WHAT'S IN GARBAGE? ARCHAEOLOGIST/GARBOLOGIST WILLIAM RATHJE, WHO DIGS DUMPS, BREAKS IT DOWN LIKE THIS:

40%   PAPER AND CARDBOARD

20%   CONSTRUCTION SCRAP AND YARD WASTE

5-9% EACH { FOOD WASTE / METAL / GLASS / PLASTIC

15%   EVERYTHING ELSE: CLOTH, RUBBER, LEATHER, ETC.

THE FIRST ADVANCE OVER THE SLOP-AND-SCAVENGE SYSTEM CAME IN THE 1800s, WITH THE CREATION OF THE

SANITARY LANDFILL.

UNFAIR

THE IDEA HERE IS TO TRUCK TRASH OUTSIDE TOWN, DUMP IT, AND COVER EACH DAY'S DEPOSIT WITH A LAYER OF EARTH.

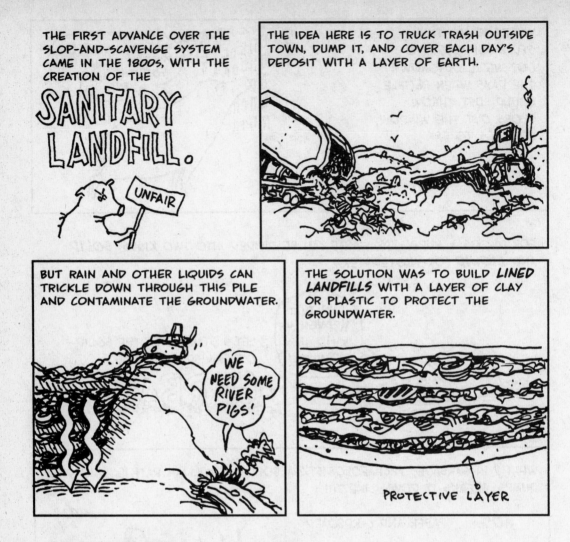

BUT RAIN AND OTHER LIQUIDS CAN TRICKLE DOWN THROUGH THIS PILE AND CONTAMINATE THE GROUNDWATER.

WE NEED SOME RIVER PIGS!

THE SOLUTION WAS TO BUILD *LINED LANDFILLS* WITH A LAYER OF CLAY OR PLASTIC TO PROTECT THE GROUNDWATER.

PROTECTIVE LAYER

LANDFILLS ARE WIDELY USED... BUT THE GARBAGE IN THEIR DEPTHS BARELY DECOMPOSES, OR DOES SO ONLY SLOWLY... SO LANDFILLS *ONLY GET BIGGER.* AS OLD ONES FILL UP, NEW SITES MUST BE FOUND, OFTEN VERY FAR FROM THE CITIES THEY SERVE.

NO, THANK YOU.

IN THE 1940s AND 1950s,
MANY CITIES TRIED TO SHRINK
GARBAGE BY BURNING IT IN

INCINERATORS.

INCINERATORS DO REDUCE
GARBAGE VOLUME, BUT THEY ALSO
PRODUCE **ASH, HEAT, CHEMI-
CAL POLLUTANTS,** AND **BAD
SMELLS.** MOST OF THE EARLY
INCINERATORS WERE SHUT DOWN
BECAUSE OF PUBLIC OPPOSITION.

IN THE 1970s AND 80s, INTEREST
REVIVED IN HIGH-TECH **WASTE-
TO-ENERGY** INCINERATORS,
WHERE HEAT FROM BURNING
GARBAGE WAS SUPPOSED TO
GENERATE ELECTRICITY. THESE
STILL HAD EMISSIONS PROBLEMS:
TO BURN CLEANLY, THEY WOULD
HAVE TO RUN CONTINUOUSLY,
REQUIRING A STEADY STREAM OF
PRE-SHREDDED GARBAGE, AND
STILL NEED EXPENSIVE
POLLUTION-CONTROL DEVICES.

BESIDES, ENVIRONMENTALISTS KEPT POINTING OUT, IF GARBAGE PRODUCTION COULD ONLY BE CUT DOWN, INCINERATORS WOULDN'T BE NEEDED IN THE FIRST PLACE.

OH, WHY DON'T YOU ENVIRONMENTALISTS RELAX AND LEARN TO EMBRACE YOUR GARBAGE?

AND HOW DO YOU CUT GARBAGE PRODUCTION?

RECYCLING

AS YOU CAN SEE FROM RATHJE'S LIST THREE PAGES BACK, A LOT OF "WASTE" CAN ACTUALLY BE REUSED AGAIN AND AGAIN. ALL IT TAKES IS SORTING THE STUFF OUT AHEAD OF TIME AND SENDING IT TO A RECYCLING PLANT—OR MAKING YOUR OWN GARDEN COMPOST—INSTEAD OF JUST THROWING IT "AWAY."

← REQUIRES ORGANIZATION

THINGS THAT CAN BE RECYCLED!

YARD WASTE
PAPER
ALUMINUM
MANY PLASTICS
WATER
BANANA PEELS
ANYTHING ELSE?

FOR RECYCLING TO WORK, INDUSTRY HAS TO HAVE THE CAPACITY TO USE THE SECOND-HAND MATERIAL. ABOUT 60% OF ALUMINUM, FOR EXAMPLE, NOW GOES BACK FOR RESMELTING, BUT AMERICAN PAPER MILLS STILL RECYCLE LITTLE. INSTEAD, RECYCLED PAPER GOES TO *EUROPE* AND *JAPAN*, WHERE TREES ARE MORE EXPENSIVE.

WHERE TREES ARE CHEAP, THEY WILL BE USED, IT SEEMS!

THE OTHER OBVIOUS WAY TO MAKE LESS GARBAGE IS TO USE **LESS DISPOSABLE STUFF.** IN THE JARGON, THIS IS CALLED

SOURCE REDUCTION.

THE MOST ENERGY-EFFICIENT
WAY TO MANAGE SOLID WASTE:
JUST DON'T MAKE ANY!!

SOURCE REDUCTION IS ABOUT **LIFESPAN:** WE NEED TO USE THINGS THAT LAST LONGER, FROM SOCKS TO HOUSES. INSTEAD OF DEBATING THE RELATIVE MERITS OF **PLASTIC** OR **PAPER** BAGS AND CUPS, TAKE YOUR OWN **CLOTH BAG** TO THE STORE... AND USE A CERAMIC CUP!

SOURCE REDUCTION AND RECYCLING PROGRAMS HAVE BEEN BIG SUCCESSES WHERE THEY'VE BEEN TRIED. IN SEATTLE, FOR EXAMPLE, PER CAPITA GARBAGE PRODUCTION FELL BY ABOUT **65%** BETWEEN 1983 AND 1993, AND FEW PEOPLE ARE TALKING ABOUT INCINERATORS ANY MORE.

FOR WATER, THE PROBLEMS ARE A LITTLE DIFFERENT: WATER WON'T JUST SIT THERE LIKE GARBAGE. IT RUNS AWAY, SPREADING WHATEVER IS IN IT ALL OVER THE PLACE.

THE FIRST PROBLEM WITH WATER IS GETTING **ENOUGH OF IT**: AMERICANS, WHOSE CITIES HAVE GOOD PLUMBING, USE ABOUT **100 GALLONS PER PERSON** EVERY DAY. IN A CITY OF MILLIONS THAT ADDS UP, AND NOW NEW YORK COMPETES WITH PHILADELPHIA FOR WATER, WHILE LOS ANGELES FIGHTS WITH THE ENTIRE STATE OF ARIZONA OVER THE SOUTHWEST'S MEAGER SUPPLY.

COLORADO RIVER

THE **100 GALLONS** BREAKS DOWN LIKE THIS:

25 GALLONS DOWN THE TOILET

20 GALLONS FOR BATHING

5 GALLONS FOR COOKING

15 GALLONS FOR DISHWASHING

20 GALLONS FOR WASHING CLOTHES

15 GALLONS FOR MISCELLANEOUS HOUSE & GARDEN USES

SOME EXCELLENT OPPORTUNITIES FOR SOURCE REDUCTION!

IN WATER-POOR CALIFORNIA, STRICT REGULATIONS HAVE CUT PER CAPITA WATER CONSUMPTION IN HALF. SOME OF THE WAYS:

* USE LOW-FLOW SHOWER HEADS

* USE TOILETS WITH SMALLER TANKS—OR PUT A BRICK IN YOUR OLD TANK

* DON'T FLUSH FOR EVERYTHING

* WATER THE GARDEN BY DRIP INSTEAD OF SPRINKLER (LESS EVAPORATION!)

* TURN OFF FAUCET WHILE BRUSHING TEETH.

MEANWHILE, AT THE OTHER END OF THE PIPE:

A LARGE CITY PUTS MILLIONS OF POUNDS OF FECES AND URINE INTO THE WATER EVERY DAY, AND DISCHARGES THEM AT A SINGLE POINT. UNTREATED WASTE WATER IS THEREFORE RICH IN ORGANIC MATTER THAT WAS ONCE (IN SOME PLACES) APPRECIATED FOR ITS FERTILIZING POWER.

DUMPED STRAIGHT INTO RIVERS AND OCEANS, HOWEVER, THIS LOAD OF ORGANIC MATTER OVERWHELMS THE ECOSYSTEM: BACTERIA AND ALGAE FLOURISH IN IT, AND AFTER A CERTAIN POINT, FISH MAY ASPHYXIATE.

THE ALGAE ARE REALLY BITING TODAY!

YET UNTREATED DUMPING IS WHAT MOST MAJOR CITIES DID UNTIL RECENTLY, WHEN THEY BEGAN BUILDING MODERN **WASTEWATER TREATMENT PLANTS.**

① ② ③

IN THESE FACILITIES, SOLIDS ARE REMOVED FROM THE WATER: FIRST TRASH IS SCREENED OUT, AND THEN THE WATER SITS IN SETTLING TANKS WHERE SLUDGE CAN SETTLE TO THE BOTTOM. FINALLY, THE WATER IS AERATED AND CHLORINATED BEFORE DISCHARGE INTO THE NEAREST WATERWAY.

THE SLUDGE IS COLLECTED--AND THEN WHAT?

THEN WE ARGUE ABOUT WHERE TO PUT IT!

IN THE PAST, SLUDGE WAS DUMPED OR INCINERATED, BUT MORE RECENTLY, AS THE U.S. HAS BANNED POURING POISONS DOWN THE DRAIN, SLUDGE HAS BECOME CLEAN ENOUGH TO BE TURNED INTO FERTILIZER PELLETS AND SOLD.

THESE PURIFICATION PLANTS DEPEND ON GOOD CITYWIDE PLUMBING TO GATHER THE RAW SEWAGE FOR TREATMENT, BUT THERE ARE MANY HUGE CITIES WITHOUT GOOD SEWERS. IN BOMBAY, CAIRO, AND MEXICO CITY, WHERE MILLIONS HAVE ONLY THE GUTTER FOR A TOILET, WASTE-DISPOSAL PROBLEMS ARE BEYOND THE SCOPE OF THIS BOOK—AND ALMOST BEYOND BELIEF.

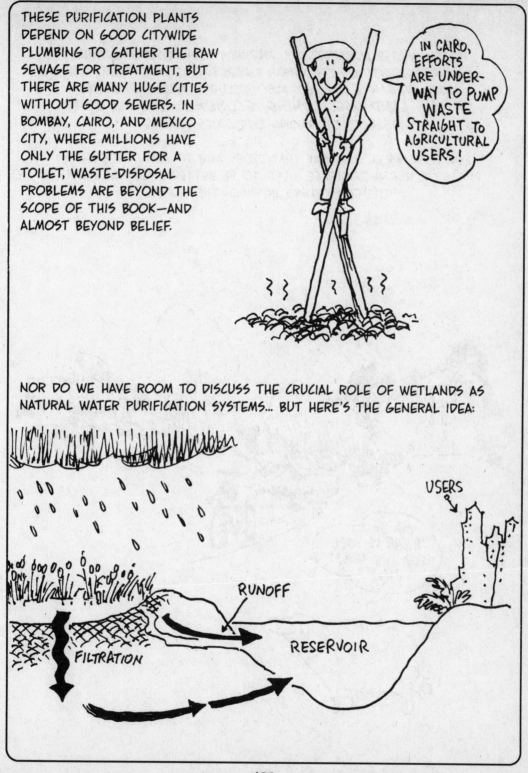

IN CAIRO, EFFORTS ARE UNDERWAY TO PUMP WASTE STRAIGHT TO AGRICULTURAL USERS!

NOR DO WE HAVE ROOM TO DISCUSS THE CRUCIAL ROLE OF WETLANDS AS NATURAL WATER PURIFICATION SYSTEMS... BUT HERE'S THE GENERAL IDEA:

USERS

RUNOFF

FILTRATION

RESERVOIR

TRAFFIC, WATER CONSUMPTION, AND WASTE DISPOSAL ARE THREE MAJOR ELEMENTS OF THE URBAN ENVIRONMENT... AND THERE ARE MORE. *URBAN ECOLOGY* ALSO INCLUDES *ARCHITECTURE, DESIGN, LAND USE, ZONING, ECONOMICS,* AND ALL OTHER ASPECTS OF MAKING OUR CITIES LIVABLE.

FOR NOW, WE'LL JUST SAY THAT CITIES AND TOWNS, LIKE OTHER PARTS OF HUMAN EXISTENCE, NEED TO BE BETTER INTEGRATED WITH THE ECOSYSTEMS IN WHICH THEY EXIST.

· CHAPTER 13 ·

POLLUTION

EVER SINCE PEOPLE BEGAN CONSUMING,
WE'VE THROWN THE LEFTOVERS AWAY...
BUT WHERE IS "AWAY?"

YOU KNOW,
AWAY!

WHEN POPULATION WAS
SMALL AND WASTES WERE
ALL ORGANIC, THIS WAS
LESS OF AN ISSUE. THE
WORLD WAS LARGE, AND
OUR LEAVINGS WERE
DILUTED OR DEGRADED
UNTIL THEY ESSENTIALLY
DISAPPEARED.

BUT THIS CHANGED WITH THE
INDUSTRIAL REVOLUTION, WHEN
POLLUTION BECAME A
BIG PROBLEM.

A PROBLEM?
THROW IT
AWAY!!

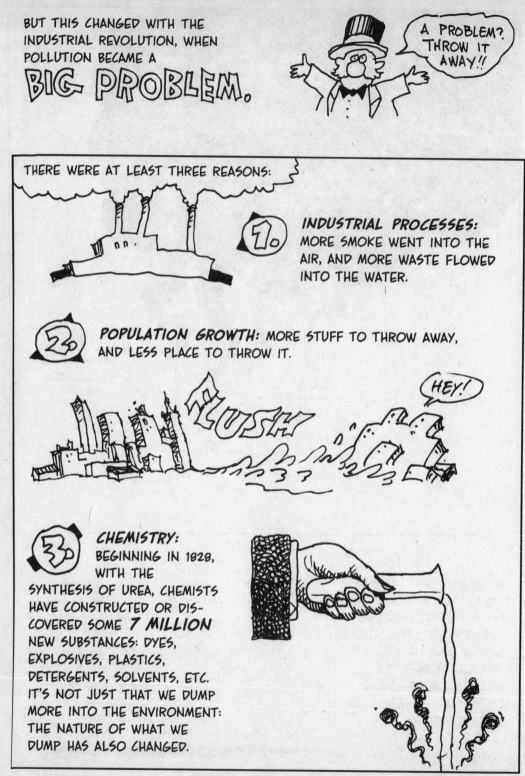

THERE WERE AT LEAST THREE REASONS:

1. INDUSTRIAL PROCESSES:
MORE SMOKE WENT INTO THE
AIR, AND MORE WASTE FLOWED
INTO THE WATER.

2. POPULATION GROWTH: MORE STUFF TO THROW AWAY,
AND LESS PLACE TO THROW IT.

HEY!

FLUSH

3. CHEMISTRY:
BEGINNING IN 1828,
WITH THE
SYNTHESIS OF UREA, CHEMISTS
HAVE CONSTRUCTED OR DIS-
COVERED SOME *7 MILLION*
NEW SUBSTANCES: DYES,
EXPLOSIVES, PLASTICS,
DETERGENTS, SOLVENTS, ETC.
IT'S NOT JUST THAT WE DUMP
MORE INTO THE ENVIRONMENT:
THE NATURE OF WHAT WE
DUMP HAS ALSO CHANGED.

THE FIRST INKLING THAT RELEASING NEW SUBSTANCES CAN HAVE UNEXPECTED RESULTS CAME IN THE 1950s. **NUCLEAR BOMBS,** EXPLODED ABOVEGROUND IN REMOTE LOCATIONS, CREATED **RADIOACTIVE FALLOUT** THAT WAS DETECTED IN AIR, RAIN, FOOD, SOIL, AND WATER ALL OVER THE WORLD.

WOA! THIS IS NEW!

IN 1954, **STRONTIUM-90,** A RADIOACTIVE ELEMENT SIMILAR TO CALCIUM, WAS FOUND IN COWS' MILK AND CHILDREN'S BONES EVERYWHERE ON EARTH.

THE BAD NEWS, TIMMY, IS THAT YOU'LL CARRY CANCER-PROMOTING ATOMS IN YOUR BONES FOR THE REST OF YOUR LIFE.

CLICK CLICK CLICK

THE GOOD NEWS IS YOU WON'T NEED THAT GLOW-IN-THE-DARK HALLOWEEN COSTUME, AFTER ALL!

WE'LL THROW FALLOUT "AWAY!"

BWOM

GOVERNMENTS ADDRESSED THIS PROBLEM BY AGREEING TO TEST WEAPONS DEEP UNDERGROUND, WHERE NO RADIOACTIVITY COULD REACH THE ATMOSPHERE.

ONLY THE GROUNDWATER!

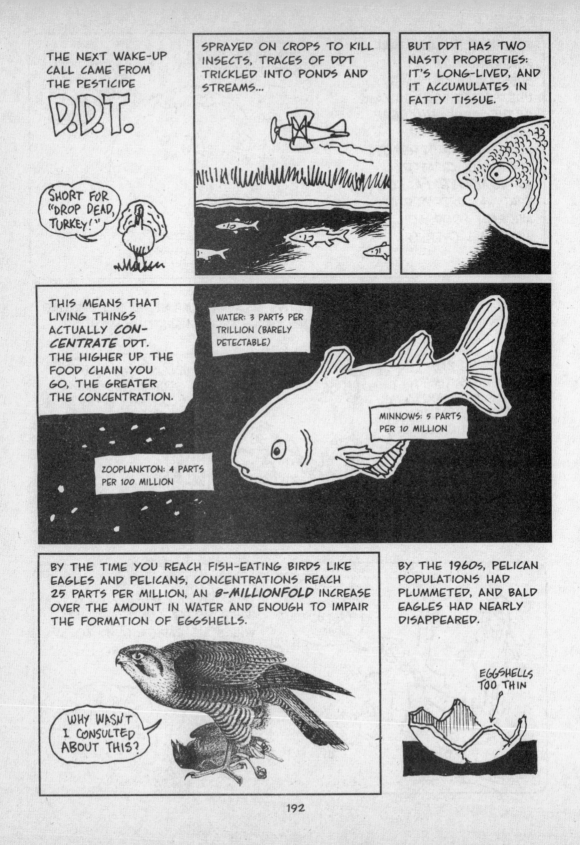

THE NEXT WAKE-UP CALL CAME FROM THE PESTICIDE **D.D.T.**

SHORT FOR "DROP DEAD, TURKEY!"

SPRAYED ON CROPS TO KILL INSECTS, TRACES OF DDT TRICKLED INTO PONDS AND STREAMS...

BUT DDT HAS TWO NASTY PROPERTIES: IT'S LONG-LIVED, AND IT ACCUMULATES IN FATTY TISSUE.

THIS MEANS THAT LIVING THINGS ACTUALLY **CONCENTRATE** DDT. THE HIGHER UP THE FOOD CHAIN YOU GO, THE GREATER THE CONCENTRATION.

WATER: 3 PARTS PER TRILLION (BARELY DETECTABLE)

MINNOWS: 5 PARTS PER 10 MILLION

ZOOPLANKTON: 4 PARTS PER 100 MILLION

BY THE TIME YOU REACH FISH-EATING BIRDS LIKE EAGLES AND PELICANS, CONCENTRATIONS REACH 25 PARTS PER MILLION, AN **8-MILLIONFOLD** INCREASE OVER THE AMOUNT IN WATER AND ENOUGH TO IMPAIR THE FORMATION OF EGGSHELLS.

WHY WASN'T I CONSULTED ABOUT THIS?

BY THE 1960s, PELICAN POPULATIONS HAD PLUMMETED, AND BALD EAGLES HAD NEARLY DISAPPEARED.

EGGSHELLS TOO THIN

192

IN 1962, SCIENCE WRITER **RACHEL CARSON** MADE AMERICA TREMBLE WITH HER BOOK *THE SILENT SPRING.*

CARSON WARNED THAT IF DDT AND OTHER PERSISTENT CHEMICALS WENT UNCHECKED, THEY WOULD SPREAD THROUGHOUT THE ENVIRONMENT, KILLING WILDLIFE AND RAISING CANCER RATES AMONG HUMANS. *CHANGE COURSE,* SHE SAID, OR RISK *POISONING THE EARTH.*

OVER THE NEXT FEW YEARS, ENVIRONMENTAL QUALITY PLUNGED... THE HUDSON RIVER DIED... THE GREAT LAKES GASPED... THE OIL-CHOKED CUYAHOGA RIVER ACTUALLY *CAUGHT FIRE...* (!)

IT'S SO ROMANTIC... MOONLIGHT IS OBSOLETE...

GRADUALLY IT DAWNED ON PEOPLE THAT YOU CAN'T THROW THINGS AWAY, BECAUSE THERE IS NO "AWAY"—AND SO BEGAN THE BIGGEST SOCIAL MOVEMENT IN U.S. HISTORY: **ENVIRON-MENTALISM.**

GREEN IS BEAUTIFUL!

NOW IF WE CAN JUST MOUNT A DIRECT-MAIL CAMPAIGN WITHOUT USING TREES...

IN 1972, THE UNITED STATES PASSED THE MOST STRINGENT POLLUTION-CONTROL REGULATIONS IN THE WORLD, AND CREATED AN

environmental protection agency

TO OVERSEE COMPLIANCE.

I'M GREEN! I'M GREEN!

NIXON

THE E.P.A. SET LIMITS ON INDUSTRIAL AND AUTOMOTIVE EMISSIONS OF VARIOUS KINDS:

TOXIC

SUBSTANCES ARE THOSE THAT ARE JUST PLAIN POISON. AT LOW DOSES, THEY ARE FATAL TO HUMANS.

WHAT? NO MORE CYANIDE DOWN THE DRAIN?

HAZARDOUS

WASTES ARE THOSE THAT MAY BURN, DISSOLVE THINGS, EXPLODE, IRRITATE, OR CAUSE ALLERGIC REACTIONS.

JUST A HINT OF SULFURIC ACID?

CARCINOGENS

INCREASE THE RISK OF CANCER. THESE INCLUDE NOT ONLY MANY CHEMICALS BUT ALSO ANYTHING THAT EMITS RADIATION.

SINCE WHEN DID A LITTLE CANCER HURT ANYBODY?

194

THE ALLOWABLE DISCHARGE LEVELS OF THESE CONTROLLED SUBSTANCES ARE BASED ON **RISK-BENEFIT** OR **COST-BENEFIT ANALYSIS,** IN WHICH THE RISKS (OR COSTS) OF POLLUTION ARE BALANCED AGAINST THE BENEFITS OF USING THE CHEMICAL.

THIS MAY SOUND REASONABLE. EVERYTHING IN LIFE IS A TRADE-OFF, RIGHT?

THE PROBLEM IS THIS: BENEFITS ARE OFTEN EASY TO SEE, BUT COSTS CAN BE HARD TO ASSESS, HIDDEN, OR ONLY APPARENT LATER ON.

PESTICIDES, FOR EXAMPLE, RAISE CROP YIELDS, LOWERING FOOD PRICES FOR EVERYONE...

BUT?

BUT PESTICIDES ARE POISONOUS. HOW DO YOU BALANCE THE BENEFIT AGAINST THE HEALTH DAMAGE TO FARMWORKERS?

SO FAR, BY USING MORE PESTICIDES!

AND WHAT DO YOU DO WHEN YOU LATER DISCOVER THAT PESTICIDE TOXINS ARE SHOWING UP IN GROUNDWATER? OR THAT INSECTS, MASTER **r**-STRATEGISTS THAT THEY ARE, BECOME **RESISTANT** TO ONE PESTICIDE AFTER ANOTHER?

POSSIBLY, JUST POSSIBLY, CONSIDER A NEW PEST CONTROL STRATEGY?

DESPITE SUCH PROBLEMS, THERE HAVE BEEN MANY ENVIRONMENTAL SUCCESS STORIES SINCE 1972. DDT WAS BANNED OUTRIGHT (FOR USE IN THE USA— AMERICAN COMPANIES CAN AND DO STILL SELL IT ABROAD!), AND BIRD POPULATIONS HAVE REBOUNDED SOMEWHAT.

BEFORE 1972, HAZARDOUS CHEMICALS WENT DOWN THE DRAIN... NOW THEY GO INTO **LICENSED, MONITORED, DOUBLE-LINED DISPOSAL FACILITIES** OR ARE INCINERATED UNDER STRINGENT EMISSION CONTROLS.

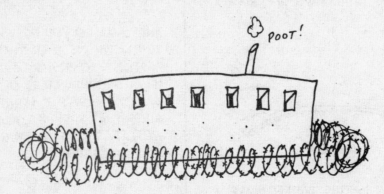

BUT ENFORCEMENT IS IMPERFECT... STANDARDS IN MANY COUNTRIES ARE WEAK... AND BEFORE 1972 WE HAD ALREADY PUMPED BILLIONS OF TONS OF CHEMICALS INTO THE ENVIRONMENT THAT WE CAN'T EASILY TAKE BACK.

WHILE TOXINS WERE A HIDDEN PROBLEM THAT SNEAKED UP ON US, *AIR POLLUTION* WAS RIGHT UNDER OUR NOSES. THE COMBUSTION OF FOSSIL FUELS IN FACTORIES AND CARS PRODUCES A HOST OF NOXIOUS STUFF: *NITROGEN OXIDES, SULFUR DIOXIDE, CARBON MONOXIDE...* AND MANY INDUSTRIAL CHEMICALS ARE GASEOUS AS WELL. THE IMMEDIATE AND OBVIOUS RESULT WAS **SMOG.**

ISN'T THIS THE SMELL OF A HEALTHY ECONOMY?

IN THE 1970S, AIR POLLUTION WAS TREATED AS A LOCAL PROBLEM: IT HAPPENED MAINLY IN CITIES. THE *CLEAN AIR ACT* IMPOSED LIMITS ON VARIOUS EMISSIONS, SO THAT CARS NOW COME WITH POLLUTION-CUTTING *CATALYTIC CONVERTORS...* WHILE POLLUTING INDUSTRIES BOUGHT SCRUBBERS, BAGHOUSE FILTERS, ELECTROSTATIC PRECIPITATORS, AND CYCLONE SEPARATORS TO CLEAN UP THEIR FACTORY SMOKE.

LOOK AT THAT! CLEAN TECHNOLOGY!

RESULT: ANOTHER (LOCAL) SUCCESS. A NEW CAR NOW EMITS ABOUT *2%* THE POLLUTION OF A 1970 MODEL... AMERICAN FACTORIES ARE CLEANER, AND MOST AMERICAN CITIES' AIR IS MORE BREATHABLE NOW THAN IT WAS 20 YEARS AGO.

BUT ATMOSPHERIC CIRCULATION IS WORLDWIDE, AND ATMOSPHERIC CHEMISTRY IS COMPLEX. THE ECOSYSTEM MOVED THOSE GASEOUS POLLUTANTS AROUND AND TRANSFORMED THEM FROM LOCAL SMOG INTO SOME REGIONAL AND EVEN GLOBAL SURPRISES.

WAK! MY FACE IS CHANGING!

ACID RAIN

IS CAUSED WHEN **SULFUR DIOXIDE** (SO_2) AND **NITROGEN OXIDES** (N_2O, NO, AND NO_2) IN THE AIR REACT WITH OTHER GASES TO FORM **SULFURIC ACID** (H_2SO_4) AND **NITRIC ACID** (HNO_3), TWO OF THE STRONGEST ACIDS IN NATURE. DISSOLVING IN RAINWATER, THESE ACIDS FALL TO EARTH.

OW!

WHEN ACID RAIN HITS THE GROUND, IT RELEASES METAL IONS FROM THE SOIL: **ALUMINUM, CADMIUM, MERCURY,** AND **LEAD,** WHICH LEACH INTO THE AQUIFER, POISONING FISH—AND ANIMALS (SUCH AS PEOPLE!) THAT EAT FISH.

GET ME ONE MILLION DOSES OF BICARBONATE OF SODA...

EVEN WITHOUT THE METALS, FISH CAN'T SURVIVE IN A STRONGLY ACID ENVIRONMENT... AND TODAY, MANY NORTHERN LAKES ARE ACIDIFIED TO SOME DEGREE.

OZONE DEPLETION

ATMOSPHERIC OXYGEN, EVERYONE'S FAVORITE GAS, NORMALLY EXISTS IN THE FORM OF A 2-ATOM MOLECULE, O_2.

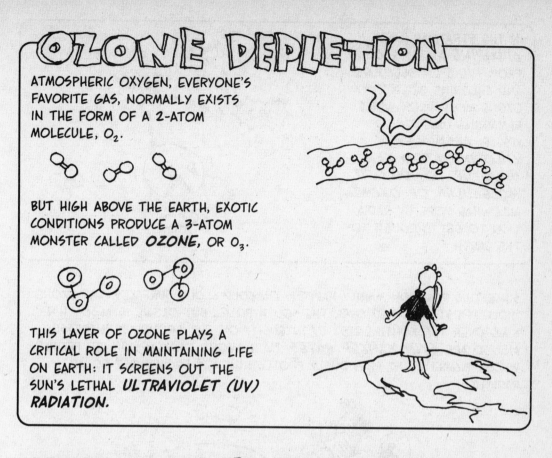

BUT HIGH ABOVE THE EARTH, EXOTIC CONDITIONS PRODUCE A 3-ATOM MONSTER CALLED **OZONE,** OR O_3.

THIS LAYER OF OZONE PLAYS A CRITICAL ROLE IN MAINTAINING LIFE ON EARTH: IT SCREENS OUT THE SUN'S LETHAL **ULTRAVIOLET (UV) RADIATION.**

THEN ALONG CAME ARTIFICIAL **CFCs...** SHORT FOR CHLOROFLUOROCARBONS. A CHEMIST'S DREAM, NON-TOXIC, NON-FLAMMABLE, HIGHLY STABLE COMPOUNDS, CFCs WERE FIRST SYNTHESIZED IN THE 1930s AND QUICKLY INCORPORATED INTO **REFRIGERATOR COILS, AEROSOL SPRAY CANS,** AND THE BUBBLES IN **STYROFOAM.** FROM THERE THEY FOUND THEIR WAY TO THE **UPPER ATMOSPHERE,** WHERE THEY **ATTACK OZONE.**

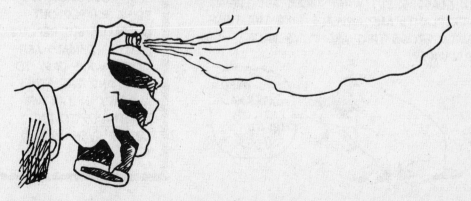

IN THE STRATOSPHERE, **CHLORINE** ATOMS ESCAPE FROM THE CFC MOLECULES... AND CHLORINE BREAKS DOWN OZONE MOLECULES, WHILE REMAINING UNAFFECTED ITSELF. *A SINGLE CHLORINE ATOM CAN UNDO UP TO 100,000 MOLECULES OF OZONE,* ALLOWING MORE UV RADIATION TO GET THROUGH TO THE EARTH.

SINCE THIS REACTION MAINLY HAPPENS ON COLD CLOUD SURFACES, THE OZONE "HOLE" FIRST APPEARED OVER THE SOUTH POLE... BUT OZONE IS ALSO THINNING OVER THE CONTINENTS... UV INTENSITY ON THE GROUND IS INCREASING, AND SO ARE **SKIN CANCER RATES**, EVEN AMONG DARKER-SKINNED PEOPLE, WHOSE PIGMENT HAD PREVIOUSLY PROTECTED THEM FROM MOST SOLAR RADIATION.

HUMANS CAN ALWAYS WEAR SUNSCREEN AND DARK GLASSES, BUT WHAT ABOUT ALL THE OTHER PLANTS AND ANIMALS, LIKE THE THIN-SKINNED FROGS THAT ARE DYING BACK WORLDWIDE?

CFC PRODUCTION IS BEING PHASED OUT GLOBALLY, BUT REPLACEMENT CHEMICALS ARE NOT ALWAYS EASY TO FIND, AND THE STUFF ALREADY IN THE AIR CONTINUES TO EAT OZONE EVEN AS YOU READ THIS.

AND FINALLY, THERE'S

GLOBAL WARMING
(PROBABLY!)

IF YOU TAKE ALL THE "POLLUTANTS" OUT OF EXHAUST GAS—THE SULFATES, NITRATES, LEAD, SOOT, ETC.—YOU ARE LEFT WITH PURE **CARBON DIOXIDE**, AND WHAT COULD BE WRONG WITH THAT? WE EXHALE THE STUFF ALL THE TIME, AND PLANTS BREATHE IT IN.

PURE PLANT FOOD!

AGAIN, THE PROBLEM IS SOLAR **RADIATION.** CARBON DIOXIDE GAS ABSORBS THE SUN'S RAYS AND WARMS UP... THAT BIT OF RADIATION IS NOT REFLECTED BACK INTO SPACE... AND SO, GENERALLY SPEAKING, THE MORE CARBON DIOXIDE IN THE AIR, THE HIGHER THE EARTH'S TEMPERATURE SHOULD BE.

THE RESULT IS LIKE HEAT TRAPPED IN A GREENHOUSE, SO THIS IS CALLED THE **GREENHOUSE EFFECT.**

EXCEPT, IN THIS "GREENHOUSE," THERE'S NO DOOR!

STUDIES OF ANCIENT EARTH CHEMISTRY SHOW THAT ATMOSPHERIC CO_2 LEVELS HAVE FLUCTUATED WIDELY OVER THE PLANET'S LONG HISTORY... AND THEY'VE BEEN STEADILY RISING FOR THE LAST CENTURY...

SCIENTISTS ARE NOW TRYING TO SEE WHETHER ANCIENT RISES IN CO_2 CORRESPOND TO PERIODS OF WARMER CLIMATE.

THERE IS MUCH DISAGREEMENT ABOUT GLOBAL WARMING... CLIMATE IS VERY IMPERFECTLY UNDERSTOOD... IT ISN'T EVEN CLEAR EXACTLY HOW TO MEASURE AVERAGE TEMPERATURES...

DO WE TRACK HIGH TEMPERATURES?

LOW TEMPERATURES?

DON'T FORGET THAT CITIES ARE WARMER BECAUSE BUILDINGS HOLD HEAT!

AND THE ALL-IMPORTANT OCEAN IS STILL MYSTERIOUS!

BUT SOMETHING SEEMS TO BE HAPPENING... SEA LEVEL HAS RISEN 6 INCHES SINCE 1900 (THE RESULT OF MELTING POLAR ICE? WATER EXPANDING FROM HEATING UP?)... HEAT WAVES ARE MORE COMMON... ALPINE PLANT ZONES ON MOUNTAIN SLOPES ARE STEADILY MOVING UPHILL, AS IF TEMPERATURES ARE BECOMING MORE TOLERABLE AT HIGHER ALTITUDES.

DO YOU REMEMBER BANANAS HERE LAST YEAR, FRITZ?

SCIENTISTS HAVE ALSO NOTICED THAT CARBON DIOXIDE IS NOT THE ONLY GREENHOUSE GAS. METHANE (CH_4), PRODUCED FROM MANY AGRICULTURAL SOURCES, FROM COW BURPS TO NATURAL FERMENTATION, MAY HAVE 20 TIMES THE EFFECT OF CO_2.

MMFF FMF?*

*"HOW DO YOU LIKE MY CATTLE-ITIC CONVERTOR?"

BY SOME ESTIMATES, WE CAN EXPECT A RISE OF **2.5°** CENTIGRADE BY THE YEAR 2050 (*4.5°* FAHRENHEIT). THIS DOESN'T SOUND BAD UNTIL YOU REALIZE THAT IN THE LAST *ICE AGE*, A *4-DEGREE DROP* PUT A *MILE OF ICE* UNDER-FOOT. THINK ABOUT 20-FOOT HIGHER SEA LEVELS... OR 120° SUMMER DAYS...

PEOPLE WILL LOOK SO RIDICULOUS, WAR WILL BE IMPOSSIBLE!

ON THE OTHER HAND, IT MAY BE THAT HIGHER TEMPERATURES WILL CREATE MORE CLOUDS, WHICH REFLECT MORE HEAT... OR PERHAPS INCREASED PLANT GROWTH WILL ABSORB THE CO_2, KEEPING US COOL.

WILL YOU MAKE UP YOUR MIND?

WHATEVER THE CASE, IT DOESN'T LOOK AS IF FOSSIL FUEL BURNING IS GOING TO SLACK OFF ANYTIME SOON... SO EXPECT CO_2 LEVELS TO KEEP RISING.

WELL, WE'LL KNOW SOON ENOUGH...

BY NOW, IT SHOULD BE CLEAR THAT WE
INCREASINGLY LIVE IN A WORLD OF OUR OWN
MAKING... AND THE QUESTION IS—WHAT ARE
WE GOING TO DO ABOUT IT???

· CHAPTER 14 ·

EARTH ISLAND

LIKE THE PEOPLE OF EASTER ISLAND, THE PEOPLE ON THE REST OF THE EARTH ARE RAPIDLY CHANGING THEIR ENVIRONMENT. ARE WE DOOMED TO A POPULATION CRASH AND A BARREN LANDSCAPE? OR CAN WE SUSTAIN A GREEN FUTURE FOR OURSELVES AND THE PLANET?

ONE OF THE SCARIEST FORECASTS OF DOOM
WAS GARRETT HARDIN'S 1968 ESSAY

"THE TRAGEDY OF THE COMMONS"

WHICH "PROVED" THAT DISASTER COULDN'T BE AVOIDED.

IMAGINE A PASTURE, GOES HIS ARGUMENT, WHERE *10 INDEPENDENT HERDERS* GRAZE THEIR CATTLE. EACH HERDER WANTS TO MAXIMIZE HIS OWN WEALTH.

IF HERDER TOM ADDS *ONE COW* TO HIS HERD, WE CALL HIS GAIN *+1.*

$$+1$$

BUT THE ENVIRONMENTAL *COST* OF AN EXTRA COW IS SHARED BY *ALL TEN HERDERS,* SO TOM'S COST IS ONLY $\frac{1}{10}$. TO TOM, *EVERY NEW COW IS PROFITABLE.*

$$-\frac{1}{10}$$

THE RESULT IS NO SURPRISE: TOM GETS AS MANY COWS AS HE CAN... AND SO DO DICK, HARRY, AND ALL THE OTHER HERDERS...

DANG. HOW'D IT GIT SO CROWDED ALL OF A SUDDEN?

UNTIL THE COMMONS IS OVERGRAZED AND TURNS INTO A DESERT. IN THIS STORY, *EVERYBODY LOSES.*

OOPS.

IN THE WIDER WORLD, THE COMMONS CONSISTS OF THE **AIR,** THE **SEA** AND ALL THE AVAILABLE RESOURCES THAT PEOPLE ARE FREE TO USE. HARDIN'S ARGUMENT MEANS WE ARE BOUND TO GOBBLE UP OUR RESOURCES AND FOUL THE PLANET IN A SHORT-SIGHTED, COMPETITIVE PURSUIT OF GAIN... INEVITABLY!!

ANOTHER CHEERFUL NOTE FROM 1968 WAS SOUNDED BY PAUL EHRLICH'S BOOK **THE POPULATION BOMB.** LIKE MALTHUS, EHRLICH SAW AN EXPLODING POPULATION AND CONCLUDED THAT MASS STARVATION WAS JUST AROUND THE CORNER.

MORE? IN 1972, **THE LIMITS TO GROWTH** (BY MEADOWS ET AL.) PREDICTED THAT POPULATION TRENDS, ENVIRONMENTAL CONDITIONS, AND THE WORLD ECONOMY WERE ALL CONVERGING TOWARD AN EARLY, POLLUTED COLLAPSE. IN THE '80s, THE GREENHOUSE EFFECT WAS DESCRIBED... OZONE DEPLETION WAS PREDICTED AND THEN OBSERVED... ETC. ETC. ETC...

SO... WHAT, IF ANYTHING, IS WRONG WITH THESE ARGUMENTS?

THERE **IS** SOMETHING WRONG, ISN'T THERE?

FIRST OF ALL, LET'S ADMIT IT: ALL THESE PESSIMISTS MAY BE RIGHT! WE HAVE THE EXAMPLE OF EASTER ISLAND, AND NATURE IS FULL OF POPULATIONS THAT HAVE COLLAPSED DOWN THE J-CURVE.

YES... IT COULD BE HOPELESS...

AT THE SAME TIME, NO MATTER HOW TEMPTED WE ARE TO LIE DOWN AND WAIT FOR DOOMSDAY, THERE ARE SOME DISTURBING SIGNS OF... HOPE!

RATS... JUST WHEN I WAS GETTING COMFORTABLE...

TAKE THE TRAGEDY OF THE COMMONS... (PLEASE!) THE FLAW THERE IS THAT HARDIN'S HERDERS **NEVER TALK TO EACH OTHER**... THERE IS NO **MANAGEMENT** OF THE COMMON LAND... THEY HAVE NO CONCEPT OF THE COMMON GOOD, NO VISION OF THE FUTURE.

YOU'RE NOTHING BUT A BUNCH OF SELFISH ATOMS!

WHY, THANK YOU, MA'AM.

IN REALITY, MANY TRADITIONAL SOCIETIES HAVE MAINTAINED COMMON RESOURCES FOR HUNDREDS, EVEN THOUSANDS, OF YEARS. THE COMMONS MAY BE OVERSEEN BY A COUNCIL OF ELDERS, AND INDIVIDUAL COMPETITION IS RESTRAINED BY CUSTOM AND RELIGION.

DON'T YOUR GODS FROWN ON UNBRIDLED GREED?

UM... THEORETICALLY...

THE INDUSTRIALIZED WORLD HAS A MIXED RECORD. IN EASTERN EUROPE, WHERE COMMUNIST IDEOLOGY STRESSED INDUSTRIAL DEVELOPMENT AT ALL COSTS, THE ENVIRONMENT SUFFERED HORRIBLY. IN THE WEST, WHERE GOVERNMENTS PAY MORE HEED TO COMPETING INTERESTS, THERE ARE LARGE NATURAL PARKS AND ENVIRON-MENTAL REGULATIONS—ALONGSIDE POLLUTION, OVERGRAZING, AND THE REST.

WELCOME TO DRIVE-IN STATE PARK!

ON A WORLDWIDE SCALE, NO COUNTRY, COMPANY, OR ORGANIZATION OWNS THE OCEANS OR ATMOSPHERE, SO AT THIS POINT IT ISN'T EASY TO SAY HOW THEY CAN BE PROTECTED.

IT'S EASY ENOUGH TO SAY HOW THEY WON'T BE PROTECTED...

UNDERSEA MINING AND PILLAGE CORPORATION

EVEN SO, THE NATIONS OF THE WORLD HAVE UNITED TO PHASE OUT OZONE-GOBBLING CFCs COMPLETELY AND TO REGULATE CERTAIN ASPECTS OF OCEAN EXPLOITATION. HOW THESE INTERNATIONAL AGREEMENTS CAN BE ENFORCED REMAINS TO BE SEEN.

AS FOR THE POPULATION BOMB—DESPITE POPULATION GROWTH, PER-CAPITA FOOD PRODUCTION HAS RISEN A LOT SINCE 1968. CHINA, INDIA, AND INDONESIA, WITH HALF THE WORLD'S POOR, BECAME SELF-SUFFICIENT IN GRAIN BY THE 1990s...

THANKS MAINLY TO HIGH-INPUT PLANT VARIETIES AND CHEMICAL FERTILIZER...

AND MODERN BIOTECHNOLOGY STILL HAS A FEW TRICKS UP ITS SLEEVE.

LIKE PLANTS AND ANIMALS THAT CONVERT FOOD ENERGY MORE EFFICIENTLY, BACTERIA THAT EAT POLLUTION, ETC...

BUT WHAT ABOUT DEPLETING OUR RESOURCES? ALTHOUGH FARMLAND IS FINITE AND FOSSIL FUELS (EXCEPT COAL) WON'T LAST LONG, ALTERNATIVE FUELS DO EXIST, AND MOST NON-FUEL RESOURCES ARE STILL PLENTIFUL. MANY MINERALS CAN BE REPLACED BY CHEAPER, MORE ENVIRONMENTALLY FRIENDLY ALTERNATIVES.

THE "RESOURCE" WE ARE SQUANDERING MOST HEEDLESSLY IS *THE BIOSPHERE ITSELF.* WE DEPEND ON PLANTS TO TURN CO_2 INTO BIOMASS AND TO FILTER POLLUTANTS FROM THE AIR AND WATER. PLANTS REGULATE THE WATER CYCLE AND PLAY A CRITICAL ROLE IN OTHER CHEMICAL CYCLES AS WELL. OTHER ORGANISMS AERATE SOILS, STORE WATER, RECYCLE PLANT NUTRIENTS, CONTROL PESTS, POLLINATE FLOWERS, AND ON AND ON AND ON!!!

OUR RECKLESS USE OF THE BIOSPHERE IS THREATENING ITS ABILITY TO DO THE JOB WE REALLY NEED: RUNNING THE CHEMICAL CYCLES AND PRESERVING THE AIR, WATER, AND SOIL.

SO WHAT IS A SUSTAINABLE ECOSYSTEM?

ONE BIG LESSON OF THIS BOOK IS THAT ECOSYSTEMS ARE **DYNAMIC**. SUSTAINABLE DOES NOT MEAN DULL AND UNCHANGING. TO BE SUSTAINABLE, AN ECOSYSTEM NEEDS TO BE RESILIENT ENOUGH TO RECOVER FROM ACCIDENTS AND TO RECOLONIZE PREVIOUSLY LOST SPACE.

IT NEEDS SOME SPRINGINESS!

SUSTAINABLE DEVELOPMENT SHOULD MEET THE NEEDS OF THE PRESENT WITHOUT DEPRIVING FUTURE GENERATIONS OF THE ABILITY TO MEET THEIR NEEDS. WE WANT TO LEAVE OUR CHILDREN A WORLD AT LEAST AS GOOD AS THE ONE WE WERE BORN INTO.

SAM, WE MAY NOT BE GETTING THAT SECOND TELEVISION, AFTER ALL...

MOM, YOU ARE **MEAN!**

IN THE PAST, WE ALWAYS ASSUMED THAT OUR INGENUITY WOULD CREATE SOLUTIONS TO PROVIDE FOR FUTURE GENERATIONS, BUT CAN THIS GO ON INDEFINITELY? AT SOME POINT, WE HAVE TO FACE THE FACT THAT WE ARE JUST A PART OF A LIMITED BIOSPHERE. THERE IS ONLY SO MUCH MATTER AND ENERGY AVAILABLE!

AS FAR AS I KNOW, I'M THE ONLY EARTH YOU'VE GOT!

212

IN THE LONG RUN, THIS CAN MEAN ONLY ONE THING: A **NO-GROWTH ECONOMY,** IN WHICH PEOPLE ENJOY A GOOD QUALITY OF LIFE BUT DON'T CONSUME SO MUCH STUFF.

WARNING: THIS IDEA MAKES SOME PEOPLE VERY UNCOMFORTABLE!

WE MAKE MONEY BY SELLING STUFF!

ZERO GROWTH IS FOR THE LONG TERM—BUT WHAT ABOUT **RIGHT NOW?**

WE'LL GROW THE ENVIRONMENTAL BUSINESS!

UNQUESTIONABLY, WE HAVE TO **REDUCE WASTE:** FARM WITHOUT EXCESS, IMPROVE EFFICIENCIES OF FUEL EXTRACTION AND ENERGY CONVERSION, AND REDUCE THE USE OF WOOD, METAL, AND PETROLEUM-BASED SYNTHETICS.

IN AGRICULTURE, WE NEED TO **INTENSIFY:** GROW MORE CROP ON THE SAME LAND (OR EVEN LESS LAND, AS CITIES ENCROACH ON FARMLAND). AGAIN THIS IS A QUESTION OF LESS WASTE AND MORE CAREFUL HUSBANDRY OF PLANT NUTRIENTS.

ALSO: PRESERVE PLANT DIVERSITY AND REDUCE MONOCROPPING!

213

POPULATION CONTINUES TO GROW, AND NEARLY HALF THE NEW MOUTHS WILL BE IN CHINA, INDIA, PAKISTAN, BANGLADESH, AND NIGERIA, ALL COUNTRIES WITH WATER PROBLEMS AND DEGRADED ECOSYSTEMS. THEIR ECOLOGICAL RECOVERY IS UNLIKELY TO COME IN THE NEXT FEW DECADES.

HOW MANY PEOPLE CAN THE PLANET SUPPORT?

WELL, I RECKON WE'RE ABOUT TO FIND OUT...

POPULATION CONTROL IS A GOVERNMENT GOAL IN MOST OF THOSE COUNTRIES. AS WE'VE SEEN, TWO OF THE MOST EFFECTIVE FORMS OF POPULATION CONTROL ARE PROSPERITY AND THE EDUCATION OF WOMEN.

IF I FELT SECURE THAT THEY'D ALL LIVE TO ADULTHOOD AND I DIDN'T DEPEND ON THEM TO SUPPORT ME IN MY OLD AGE, I MIGHT NOT HAVE SO MANY OF 'EM!

INFORMATION IS A CATALYST FOR CHANGE. AS WE ENTER THE INFORMATION AGE, IT IS BECOMING POSSIBLE TO BRING THE LATEST ADVANCES IN BIOTECHNOLOGY, AGRICULTURAL PRACTICES, IRRIGATION TECHNIQUES, AND EVEN COOKSTOVE TECHNOLOGY TO THE REMOTEST PLACE ON EARTH—AND TO LEARN FROM TRADITIONAL CULTURES HOW THEY DO IT!

HM... IT'S FULL OF BOOKS CALLED "THE CARTOON GUIDE TO THE ENVIRONMENT..."

214

FUNDAMENTALLY, GLOBAL CHANGE WILL COME FROM THE *5+ BILLION INDI- VIDUALS* WHO LIVE ON THIS PLANET. USING LESS PAPER, METAL, AND PLASTIC... GROWING AND BUYING PESTICIDE-FREE FOOD... PLANTING TREES... CONSERVING WATER... USING LONGER-LIVED GOODS... RECYCLING AND BUYING RECYCLED PRODUCTS... HAVING FEWER CHILDREN... AND PUSHING OUR LEADERS TOWARD SUSTAINABLE POLICIES: ALL THESE ARE ELEMENTS OF GLOBAL CHANGE.

WE ARE ALL PART OF THE BIOSPHERE ON THIS REMARKABLE ISLAND OF LIFE. EVERY INDIVIDUAL WHO "THINKS GLOBALLY AND ACTS LOCALLY" LIKE A RESPONSIBLE HETEROTROPH IS PART OF THE MOVEMENT TOWARD SUSTAINABILITY.

DON'T FORGET TO TURN OUT THE LIGHTS!

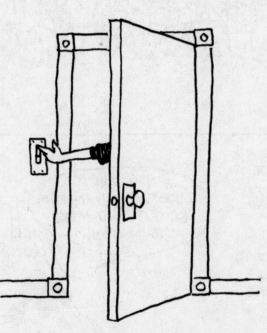

NOW IF
ONLY BOOKS
WEREN'T MADE
OF TREES...

BIBLIOGRAPHY

BAHN, PAUL AND JOHN FLENLEY. *EASTER ISLAND, EARTH ISLAND.* NEW YORK: THAMES AND HUDSON, 1992. 200 YEARS OF RESEARCH ON EASTER ISLAND, AND READABLE TOO.

CARTER, VERNON, AND TOM DALE. *TOPSOIL AND CIVILIZATION.* NORMAN, OKLAHOMA: UNIVERSITY OF OKLAHOMA PRESS, 1974. A CLASSIC ON DIRT.

CARTWRIGHT, FREDERICK. *DISEASE AND HISTORY.* NEW YORK: THOMAS Y. CROWELL, COMPANY, 1972.

DAY, DAVID. *THE DOOMSDAY BOOK OF ANIMALS: AN ILLUSTRATED ACCOUNT OF THE FASCINATING CREATURES WHICH THE WORLD WILL NEVER SEE AGAIN.* NEW YORK: VIKING PRESS, 1981. FROM BRITISH MUSEUM FILES, QUIRKY AND BEAUTIFULLY ILLUSTRATED.

DEBEIR, JEAN-CLAUDE, JEAN-PAUL DELEAGE AND DANIEL HEMERY. *IN THE SERVITUDE OF POWER: ENERGY AND CIVILIZATION THROUGH THE AGES.* LONDON: ZED BOOKS, 1991. UNBELIEVABLY ERUDITE, WITH A GREAT VIEW OF SLAVE SYSTEMS.

DEEVEY, E.S. "MAYAN URBANISM: IMPACT ON A TROPICAL KARST ENVIRONMENT." *SCIENCE.* OCT. 19, 1979: 298-306. MANY THEORIES COMPETE, BUT THOSE PHOSPHORUS SPIKES IN THE SEDIMENT CORES ARE MOST PERSUASIVE!

EHRLICH, PAUL R. AND ANNE H. EHRLICH. *HEALING THE PLANET: STRATEGIES FOR RESOLVING THE ENVIRON-MENTAL CRISIS.* READING, MA.: ADDISON-WESLEY PUBLISHING COMPANY, 1991.

GIRARDET, HERBERT. *THE GAIA ATLAS OF CITIES: NEW DIRECTIONS FOR SUSTAINABLE URBAN LIVING.* NEW YORK: ANCHOR PRESS, DOUBLEDAY, 1992. WORLDWIDE VIEW OF CITY SYSTEMS.

HARDIN, GARRETT. "THE TRAGEDY OF THE COMMONS." *SCIENCE.* DEC. 13, 1968: (DON'T KNOW WHICH PAGES. SORRY!) SHORT, BUT IT MADE A BIG STIR.

HIESER, CHARLES B. *SEED TO CIVILIZATION: THE STORY OF FOOD.* CAMBRIDGE, MA.: HARVARD UNIVERSITY PRESS, 1990. GREAT STUFF IF YOU'VE NEVER SEEN IT, OTHERWISE ZZZZZZ.

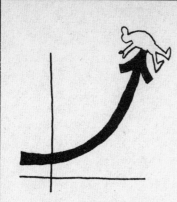

LOVELOCK, J.E. *GAIA: A NEW LOOK AT LIFE ON EARTH.* NEW YORK: OXFORD UNIVERSITY PRESS, 1979. BRIEF AND PROVOCATIVE.

MCNEILL, WILLIAM. *PLAGUES AND PEOPLES.* NEW YORK: ANCHOR BOOKS, 1976. NOT WELL WRITTEN, BUT THOUGHT-PROVOKING.

MEADOWS, D.H., D.L. MEADOWS AND J. RANDERS. *BEYOND THE LIMITS: CONFRONTING GLOBAL COLLAPSE, ENVISIONING A SUSTAINABLE FUTURE.* POST MILLS, VT.: CHELSEA GREEN PUBLISHING COMPANY, 1992. THE SEQUEL TO LIMITS TO GROWTH WITH LOTS OF DATA.

MEADOWS, D., ET AL. *LIMITS TO GROWTH.* NEW YORK: UNIVERSE BOOKS, 1972.

MILLER, G. TYLER. *LIVING IN THE ENVIRONMENT: SEVENTH EDITION.* BELMONT, CA.: WADSWORTH, 1992. FOR A COLLEGE TEXTBOOK ON ENVIRONMENTAL SCIENCE, IT CONVEYS A SURPRISING LEVEL OF FEELING ALONG WITH THE FACTS.

MYERS, NORMAN, ED. *GAIA: AN ATLAS OF PLANET MANAGEMENT.* NEW YORK: ANCHOR PRESS, DOUBLEDAY, 1984. A GOOD SYSTEMS OVERVIEW, BUT SOMETIMES OVERWHELMING GRAPHICS.

NEWMAN, LUCILE. *HUNGER IN HISTORY: FOOD SHORTAGE, POVERTY AND DEPRIVATION.* CAMBRIDGE: BASIL BLACKWELL, INC., 1990. LOTS OF SKELETAL STUDIES AND OLD FAMINE RECORDS.

PONTING, CLIVE. *A GREEN HISTORY OF THE WORLD: THE ENVIRONMENT AND THE COLLAPSE OF GREAT CIVILIZATIONS.* NEW YORK: ST. MARTIN'S PRESS, 1991. A BRILLIANT, DETAILED VIEW OF HUMAN IMPACTS ON EARTH.

RATHJE, WILLIAM AND CULLEN MURPHY. *RUBBISH! THE ARCHEOLOGY OF GARBAGE.* NEW YORK: HARPER PERENNIAL, 1992. LANDFILL DISSECTIONS AND GARBAGE REFLECTIONS.

WILSON, E.O. *THE DIVERSITY OF LIFE.* NEW YORK: W.W. NORTON AND COMPANY, 1992. EVERYTHING ABOUT BIODIVERSITY BY A MASTER.

ZINSSER, HANS. *RATS, LICE, AND HISTORY.* NEW YORK: BLUE RIBBON BOOKS, 1935. ECCENTRIC, ERUDITE, AND VERY FUNNY IN AN ODD, DRY WAY, THIS CLASSIC EXPLORES THE HISTORY OF TYPHUS.

ZISWILER, VINZENZ. *EXTINCT AND VANISHING ANIMALS: A BIOLOGY OF EXTINCTION AND SURVIVAL.* NEW YORK: SPRINGER-VERLAG, 1967. GREAT DOCUMENTATION OF THE FORMER TRADE IN ANIMAL PARTS.

◆ INDEX ◆

India, 165, 167
  food production in, 210
  population growth in, 214
Indicator species, 133
Indonesia, 124–25, 210
Industrial revolution, 141–45, 190
Inertia, ecological, 62
Infectious diseases, 110, 112–18
Influenza, 116
Inland wetlands, 45
Insects, 46, 49, 50
  in rain forest, 54–55
  species of, 122
Insulation, 83, 153
Internal combustion engine, 145, 147
Iodine, 19
Iraq, 95
Iron, 19
Island ecosystems, 48, 123–24, 132, 205. *See also* Easter Island.
Isolation, speciation and, 29
Ivory, 126

J

Jackrabbits, 52
Javan tiger, 133
Jays, 134
J curve, 107, 113
Jute, 165

K

Kangaroos, 123
Keystone species, 130–31
Kilns, 140
K selection, 32–33, 36, 38

L

Labor, division of, 89
Lakes, zones of, 44
Lampreys, 79
Landfills, 179–80
Landlord class, 102
Laterites, 55
Lava, 58, 59
Leaves, soil and, 51
Leeches, 77
Legumes, 91, 92
Lichen, 46, 51, 59
Light bulbs, 145, 147
Limiting factors
  adaptability as, 31, 76, 80
  diseases as, 112–18
  famine as, 111–12

for humans, 98–99, 108–19
  for predators, 80
  war as, 109–10
Limits to Growth, 207
Limnetic zone, 44
Littoral zone, 44
Llamas, 91
Logging, 50, 60, 95
Los Angeles, 184
Lotka-Volterra graph, 78
Love, Charles, 6
Lovelock, James, 16–18
Lubricants, 153
Lungfish, 46
Lynx, 49, 50

M

Macroconsumers, 73
Macronutrients, 19
Madagascar, 123–24
Maize, 91
Malthus, Thomas, 108–9, 207
Mammals
  Australian, 123, 125
  species of, 122
Mammoths, 85
Mangoes, 165
Mangrove swamps, 42
Manure, 92
Margulis, Lynn, 16–18
Marshes, 42, 45
Mass transit, 152, 153, 177
Mayans, 96
Meadows, D., 207
Meat consumption, 171–72
Mechanical energy, 139, 143
Medicinal plants, 136
Mediterranean biome, 48
Megadiversity, 123
Melons, 165
Metalworking, 140
Methane, 202
Mexico, 91, 96, 165, 187
Mice. *See* Rodents.
Microconsumers, 73
Micronutrients, 19
Migrant workers, 162
Migration routes, 132, 134
Millet, 91, 166
Millipedes, 73
Mink, 50
Mojave desert, 53
Moose, 50
Mosquitos, 67, 122
Moss, 46, 51, 59

ABOUT THE AUTHORS

ALICE OUTWATER IS AN ENVIRONMENTAL ENGINEER TURNED ECOLOGY WRITER. A GRADUATE OF THE UNIVERSITY OF VERMONT WITH AN M.S. FROM MIT, SHE HELPED MANAGE THE SLUDGE FOR THE SIX-BILLION-DOLLAR BOSTON HARBOR CLEAN-UP. THIS IMMERSION IN THE OTHER END OF THE TOILET LED TO NUMEROUS TECHNICAL PAPERS AND A TEXTBOOK, *REUSE OF SLUDGE AND MINOR WASTEWATER RESIDUALS.* SHE LIVES WITH HER HUSBAND BOB AND THEIR SON SAM ON A FARM IN VERMONT, WHERE SHE WRITES AND CONSULTS IN WASTEWATER MANAGEMENT. HER THIRD BOOK, *WATER: A NATURAL HISTORY*, WILL BE PUBLISHED BY HARPERCOLLINS IN 1996.

LARRY GONICK IS THE AUTHOR OR COAUTHOR OF NUMEROUS WORKS OF GRAPHIC NON-FICTION. TWO OF HIS BOOKS, *THE CARTOON HISTORY OF THE UNIVERSE* AND *THE CARTOON GUIDE TO PHYSICS*, HAVE BEEN ADAPTED TO CD-ROM. A CONTRIBUTING EDITOR OF *DISCOVER* MAGAZINE WITH THE BIMONTHLY FEATURE "SCIENCE CLASSICS," HE SPENT THE 1994-95 ACADEMIC YEAR AS A KNIGHT SCIENCE JOURNALISM FELLOW AT MIT. HE LIVES IN SAN FRANCISCO WITH HIS WIFE LISA AND HIS DAUGHTERS SOPHIE AND ANNA. FOR THEIR SAKE, HE HOPES YOU WILL ALL HELP TO SAVE THE ENVIRONMENT!